Process Science and Engineering for Water and Wastewater Treatment

Series Editor

Professor Tom Stephenson

Volume Editor

Dr. Simon Judd

IWA Publishing

Published by IWA Publishing, Alliance House, 12 Caxton Street, London SW1H 0QS, UK

Telephone: +44 (0) 20 7654 5500; Fax: +44 (0) 20 7654 5555; Email: publications@iwap.co.uk
Web: www.iwapublishing.com

First published 2002

Printed by TJ International (Ltd), Padstow, Cornwall, UK

British Library Cataloguing in Publication Data
A CIP catalogue record for this book is available from the British Library

Library of Congress Cataloging-in-Publication Data
A catalog record for this book is available from the Library of Congress

ISBN: 1 900222 75 2

Contents

Introduction to the *Water and Wastewater Process Technologies* Series

Unit operations are linked together in a flowsheet to provide water and wastewater treatment.

This series of texts, termed *modules*, is designed to provide an education in water and wastewater treatment from a process engineering perspective. At the end, you should have a thorough understanding of the design, operation and management of water and wastewater treatment processes. This might be the treatment of an industrial effluent to a standard acceptable for disposal to sewer, or treatment of municipal sewage to meet an environmental discharge consent, or the production of water suitable for drinking abstracted from a river. In almost all cases, a single "process", usually termed a *unit operation*, will not be able to achieve the required level of treatment. Treatment is achieved through linking the right unit operations in a *flowsheet*.

With the emphasis on the public utility infrastructure and environmental impacts, rather than the processes, technologies required for provision of potable quality water and those required for treatment of wastewaters have been taught separately. Teaching tends to follow an "application down" approach. Elements of training at almost all levels of the water sector are usually presented in this way. For example, a programme on wastewater treatment might be structured with courses entitled *Introduction to municipal wastewater treatment; Advanced municipal wastewater treatment; Low cost sewage treatment; Industrial effluent treatment;* and so on. For drinking water, it is likely to be an *Introduction to potable water treatment*, followed by *Advanced potable water treatment*, etc. This means that unit operations relying upon different scientific principles are taught together, arguably making it difficult to instil a true understanding of the principles governing the design, operation and management of the unit operations.

The "application down" approach looks even more flawed when we consider today's needs for different qualities of water other than potable supply or municipal sewage treatment, in addition to ever more stringent quality standards. Examples might include provision of ultra-pure water for microchip production, water suitable for crop irrigation, in-building water recycling for non-potable uses, potable supply in rural areas, water for manufacturing and treatment of industrial effluents, particularly when high in inorganic content. Very different processes may be needed to provide solutions in these situations.

When designing, operating or managing a process to provide a certain quality water, it is better to first consider the individual unit operations that when linked together form the required process flowsheet for the application. This is a chemical engineering approach, as it is about the conception, development and exploitation of processes and their products. The process could be water or effluent treatment; the product a less polluted aqueous stream.

Understand water and wastewater treatment through a chemical engineering approach.

Chemical engineers use physical, chemical and biological sciences and mathematics to provide a systems approach to the understanding of changes which take place in processes, from the molecular to global scale, and to establish methods which can be employed to achieve required changes in composition, energy content, structure or physical state. In other words, chemical engineering uses a "science up" approach to solving process problems.

This series of modules teaches the removal of pollutants from water based on just such a "science up", chemical engineering approach. To understand a unit operation, the fundamental biology, chemistry or physics underlying that unit operation needs to be properly understood. Nearly all unit operations are applicable to different levels of water and wastewater treatment, e.g. ultrafiltration membrane processes can be used in flowsheets for treating municipal and industrial effluents, as well as potable supply and pure water. The fundamental concepts of transmembrane pressure, flux rates, fouling etc, remain the same, whatever the application. Therefore the texts in this series are organised on just such a "science up" basis.

The first module, *Process Science and Engineering*, covers the chemistry, biology and chemical engineering required for a proper understanding of the unit operations covered in the main technical modules. The second module, *Principles of Water and Wastewater Treatment Processes*, introduces these unit operations, describing their overall form, performance and application. These two modules provide an introduction to the remaining five modules that describe the unit operations based upon their scientific principles: *Biological Processes, Chemical Processes, Physical Processes, Membrane Technology,* and *Sludge Treatment, Management and Utilisation*.

SERIES MODULES
- Process Science and Engineering
- Principles of Water and Wastewater Treatment Processes
- Biological Processes
- Chemical Processes
- Physical Processes
- Membrane Technology, and
- Sludge Treatment, Management and Utilisation

School of Water Sciences at Cranfield University

Cranfield UNIVERSITY

The series of texts on *Water and Wastewater Process Technologies* has been developed from the modular Masters programmes delivered by the School of Water Sciences at Cranfield University.

The School of Water Sciences specialises in treatment, use and management of water.

The School of Water Sciences undertakes research and education in technology and associated scientific, engineering and policy issues for the treatment, use and management of water. It is based on the Cranfield campus of Cranfield University, an institution unique in UK Higher Education. The University was founded as the Royal College of Aeronautics in 1946 and now occupies three campuses in the south of England. The Cranfield campus is approximately 80 km north of London and is the largest postgraduate-only campus in the UK. In addition to a fully operational airfield, it has its own sewage treatment works. Here the School of Water Sciences has a pilot-plant facility. In addition there are state-of-the-art laboratories for undertaking research and teaching in water and wastewater processing on the campus. These include modern analytical instrumentation facilities, chemistry and microbiology laboratories and experimental rig areas

Water Sciences at Cranfield offers a comprehensive programme of Masters and Doctoral research degrees.

Research programmes offered include the 3 year full-time Doctorate of Philosophy (PhD) and the 4 year Engineering Doctorate (EngD). The latter combines an in-depth technical thesis with a high-level management programme. Research projects always involve support from the water industry, often financial as well as technical. Research themes include municipal water and wastewater treatment, industrial effluent treatment, environmental monitoring and modelling, groundwater protection, pollution control, odour treatment and management, waste management, technology policy, water and society, water reuse and biosolids disposal. All research students are encouraged to publish in scientific journals during their studies and are given the opportunity to present their research at international meetings. Each student is assigned an academic member of staff who supervises the researcher throughout their project.

Masters Programmes in
Water Pollution Control Technology
Water and Wastewater Technology
Water and Wastewater Engineering

The teaching programme of the School centres on the full-time (*Water Pollution Control Technology, Water and Wastewater Engineering*) and part-time (*Water and Wastewater Technology/Engineering*) MSc courses. All of these are modular in structure based upon the texts in the series *Water and Wastewater Process Technologies.*

The full-time MSc is available as one year of intensive study. The course consists of ten taught modules, a series of design projects undertaken in teams and a thesis project. A Postgraduate Diploma (PgD) can be taken over 7 months and includes 10 taught modules and the group design projects. A Postgraduate Certificate (PgC) can be taken over 4 months and includes 6 taught modules along with a design project.

The part-time MSc can be taken over 2 to 5 years whilst in employment.

The taught modules include the technical subjects that form the series *Water and Wastewater Process Technologies* as well as modules on technical management subjects such as project management and reliability engineering and risk management.

During design projects, the students work in multi-disciplinary teams to produce detailed process flowsheets for a variety of treatment problems. Concepts learned during the taught modules are put into practice. Designs include advanced potable water treatment, municipal wastewater treatment and industrial effluent processes.

The full-time MSc thesis projects are undertaken in Water Sciences' modern facilities on the campus, or more often off-site in co-operation with international and UK water companies. Thesis projects have been sponsored by many of the world's largest water companies, including all ten privatised water and sewerage utilities in England and Wales. There is also the opportunity to undertake the thesis project at other universities across the globe. Part-time students usually undertake thesis projects with their employer.

Students from the UK, Europe and the rest of the world attend the full-time MSc courses. Countries represented include Argentina, Brazil, Columbia, France, Mexico, Norway, Italy, Portugal, Spain, Syria, Taiwan and Thailand. The undergraduate qualifications of students have included degrees and diplomas in many science and engineering subjects. These have been chemical, civil and mechanical engineering, biological sciences, chemistry, environmental sciences and many other subjects.

The School of Water Sciences organises conferences in key areas of technology appropriate to the water sector. Generally, each conference includes presentations by leading international figures working in these specialist fields. These presenters are drawn from industry as well as academia. Delegates receive sets of papers that are also published following the close of the conference. These events are often held in conjunction with other bodies such as the IWA, and may include a small exhibition organised parallel to the conference so that delegates can meet with personnel from companies marketing associated products and services.

Cranfield UNIVERSITY

For further details on Masters level courses, research and conferences run by the School of Water Sciences please see http://www.cranfield.ac.uk/sims/water/
or
Email: watersciences@cranfield.ac.uk

or write to:
The School of Water Sciences, Cranfield University, Cranfield, Bedfordshire, MK43 0AL, UK

Tel: +44 (0)1234 750111
Fax: +44 (0)1234 751671

Editors

Series Editor

Professor Tom Stephenson is Professor of Water Sciences at Cranfield University. He graduated with a BSc in Biochemistry from the University of York, UK, and has a PhD in Civil Engineering from Imperial College, University of London. After a period as a post-doctoral researcher at Imperial, he was appointed Senior Lecturer in Chemical Engineering at Teesside Polytechnic, UK. He joined the School of Water Sciences at Cranfield University in 1990 and became Professor of Water Sciences in 1994. Professor Stephenson is a Chartered Engineer and a Fellow of the Institution of Chemical Engineers and the Chartered Institution of Water and Environmental Management.

Professor Stephenson has developed both full-time and part-time Masters programmes in water and wastewater process technologies at Cranfield. His research interests are centred on wastewater treatment, including biological, chemical and physical processes. Professor Stephenson has consulted for many national and international companies, contributed to government and OECD Reports and has led UK government overseas science and technology missions. He is also a Director of British Water, the leading UK Trade Association for the water industry in the home and international markets. Professor Stephenson participates as regularly as possible in activities endorsed by the Campaign for Real Ale (CAMRA).

Volume Editor

Dr. Simon Judd is Reader in Water Sciences. He has a BSc in Chemistry from the University Bath, UK, and an MSc in Electrochemical Science from Southampton University, UK. Dr Judd took a PhD on electrophoretic depth filtration while in the School of Water Sciences at Cranfield. Dr. Judd subsequently worked in the Chemical Process Development Department at AEA Winfrith, UK, where he continued work on filtration as well as other environmental-related processing. Rejoining Cranfield in 1992, Dr. Judd was promoted to Reader in Water Sciences in 1999. Dr Judd is a chartered chemist and a member of the Royal Society of Chemistry. His main areas of research interest are membrane and chemical processes for water and wastewater treatment, swimming pool water treatment and industrial water recycling technologies.

The Series Editor and Volume Editor acknowledge the technical writing contribution of Mr Richard Hill, Visiting Fellow in Water Sciences.

How to Use This Book

This text, or module, has been designed for *individual self-paced study* i.e. you can learn in your own time without requiring additional face-to face tuition.

Each section of the main text gives step-by-step learning in a particular subject. These sections are termed *Units. Aims and objectives* for each Unit are given at the start. Following these, essential prerequisites are highlighted, i.e. which other Units contain concepts that you should understand before starting on the particular Unit.

At the start of each Section, in the margin you will find an hourglass symbol with an approximation of how long you will need to spend on that Section. This is only an approximation, as this will be dependent on your prior knowledge and how easily you understand the subject.

This section requires a study time of about 1 hour

The main text of each Section is enhanced with equations, figures and tables to aid understanding. As you work through each section you will find, in the right hand margin, *key points* highlighted in boxes.

KEY POINTS
As you work through each section you will find, in the right hand margin, key points highlighted in a box

In the right hand margin there are also *exercises* to help understand the material in the main text. You should attempt all exercises as you go through the text and then check your answers. These are provided at the end of each section.

Ex 1.1
In the margin there are also exercises to help you understand the material in the main text.

At the end of the main text for each unit there are *self-assessment questions* that you should attempt to test your understanding. The answers, with working, are given at the end of the text. It is important that you do attempt the self-assessment questions *before* you look at the answers. This will help your understanding of the subject.

Introduction to Process Science and Engineering for Water and Wastewater Technologies

Process engineering depends on the fundamental sciences of chemistry, physics and biology. You can use the printed material to learn about the key fundamental, underpinning chemistry, biology and engineering of water and wastewater treatment processes in your own time.

The text is divided into the following Units:

1. **Fundamentals of Water Chemistry**
2. **Chemical Kinetics and Equilibria**
3. **Colloid and Surface Chemistry**
4. **Fundamentals of Microbiology**
5. **Fundamentals of Biochemistry**
6. **Microbial Kinetics**
7. **Fundamentals of Process Engineering**
8. **Mass and Heat Balances**
9. **Introductory Mass and Heat Transfer**
10. **Reactor Design Theory**
11. **Engineering Hydraulics**
12. **Particle Settlement**

The subjects involved are chemistry, biology and engineering. For each subject there are a number of Units: 3 for chemistry (Units 1–3), 3 for biology (Units 4–6) and 6 for engineering (Units 7–12).

Process Science and Engineering for Water and Wastewater Technologies covers the subjects you will need to assimilate so that you can fully understand the unit operations described in later modules.

While every attempt has been made to ensure that all the information is correct, it is almost inevitable that there will be some errors in a document of this size. If you do find any, please let us know.

1. Fundamentals of Water Chemistry

Contents

Aims and Objectives

This Unit introduces the basic concepts of chemistry as applied to water and wastewater treatment.

After studying this Unit you should be able to:

1. explain in your own words the following terms and concepts:

   atom, molecule and ion
   salt, acid and base
   molecular weight
   equivalent weight
   concentration
   conductivity
   stoichiometry

2. carry out calculations for:

   balancing equations
   converting concentration units

It is not essential for you to memorise any equations or constants.

It is important that you are able to correctly complete all the self assessment questions at the end of these notes.

Essential Prerequisites

It is not necessary to have completed any other Units before undertaking this Unit.

1.1 Introduction

In any water treatment process where dissolved or suspended pollutants are removed, it is likely that their removal in some way or other involves chemistry. Even a straightforward apparently physical process such as filtration involves some chemistry, because the capture of particles depends to some extent on the chemistry of the sand grain surface. In many other important processes, including ion exchange, reverse osmosis and coagulation-flocculation to name but a few, it is simply impossible to ignore chemical phenomena or avoid calculations founded on basic chemical principles. Chemistry not only determines how some processes work on a fundamental level, but is also used to identify both the required process or processes and the operating conditions.

Although conceptually simple, involving nothing more than the addition and mixing of chemicals, chemical methods are none-the-less extremely complicated on a microscopic level and sensitive to the chemical conditions of the water. Efficient operation of these processes therefore relies on knowledge of the water chemistry and, if possible, its close control. Specifically, it is normally crucial to ensure that the solution pH is kept within specific limits in order to ensure that the required chemical reaction proceeds as quickly as possible. It is for this reason that precipitation, coagulation/flocculation and chlorination unit processes are usually preceded by pH adjustment. Finally, it has recently been demonstrated that, not withstanding the abstruse nature of the solution chemistry, it is none-the-less theoretically possible to model chemical processes from a consideration of fundamental equilibrium thermodynamics. Computer models based on this approach have already been developed and successfully employed for predicting process performance in the specific area of wastewater precipitation.

This Unit is a basic, step-by-step guide to chemistry for the purposes of water treatment process design. These notes summarise the more important facts, some of which are simplifications, but accurate enough for the purposes of following discussions. Chemistry at this level is easy, but it uses a number of terms and many jargon words which may be unfamiliar. These have to be learnt, and so to help identify them, they are *italicised* in the text and a brief glossary of terms is provided.

Section 1.2 requires a study time of about 2 hours

ATOM
An atom is the smallest particle of an element which can take part in a chemical reaction.

PERIODIC TABLE
The Periodic Table identifies groups of elements with similar chemical properties.

1.2 Chemical Building Blocks

1.2.1 Atoms

Matter is made of *atoms*, which are indestructible under normal conditions. A material made up of only one type of atom is called an *element*. The names and corresponding chemical symbols – in effect, shorthand notation – of the elements of interest in water treatment are listed in Table 1.1. The table includes data for the *atomic weight*, or the relative weight of atoms of the elements.

Table 1.1 – Common elements in water treatment

Name	Symbol	Atomic Weight
Aluminium	Al	27
Calcium	Ca	40
Carbon	C	12
Chlorine	Cl	35.5
Hydrogen	**H**	**1**
Iron	Fe	56
Magnesium	Mg	24
Manganese	Mn	55
Nitrogen	N	14
Oxygen	O	16
Silicon	Si	28
Sodium	Na	23
Sulphur	S	32

You should note that Table 1.1 does not represent a comprehensive list of all the elements. In fact, there are 109 elements currently known to man. 103 of these are very often presented in the form of the *Periodic Table* (Figure 1.1).

Period	Group Ia	Group IIa	Group IIIa	Group IVa	Group Va	Group VIa	Group VIIa	Group VIII			Group Ib	Group IIb	Group IIIb	Group IVb	Group Vb	Group VIb	Group VIIb	Group O
1 1s	1 H																1 H	2 He
2 2s2p	3 Li	4 Be											5 B	6 C	7 N	8 O	9 F	10 Ne
3 3s3p	11 Na	12 Mg											13 Al	14 Si	15 P	16 S	17 Cl	18 Ar
4 4s3d 4p	19 K	20 Ca	21 Sc	22 Ti	23 V	24 Cr	25 Mn	26 Fe	27 Co	28 Ni	29 Cu	30 Zn	31 Ga	32 Ge	33 As	34 Se	35 Br	36 Kr
5 5s4d 5p	37 Rb	38 Sr	39 Y	40 Zr	41 Nb	42 Mo	43 Tc	44 Ru	45 Rh	46 Pd	47 Ag	48 Cd	49 In	50 Sn	51 Sb	52 Te	53 I	54 Xe
6 6s (4f) 5d 6p	55 Cs	56 Ba	57* La	72 Hf	73 Ta	74 W	75 Re	76 Os	77 Ir	78 Pt	79 Au	80 Hg	81 Ti	82 Pb	83 Bi	84 Po	85 At	86 Rn
7 7s (5f) 6d	87 Fr	88 Ra	89** Ac															

*Lanthanide series 4f	58 Ce	59 Pr	60 Nd	61 Pm	62 Sm	63 Eu	64 Gd	65 Tb	66 Dy	67 Ho	68 Er	69 Tm	70 Yb	71 Lu
**Actinide series 5f	90 Th	91 Pa	92 U	93 Np	94 Pu	95 Am	96 Cm	97 Bk	98 Cf	99 Es	100 Fm	101 Md	102 No	103 Lr

Fig. 1.1 – Periodic table of the elements

This table presents the elements sequentially in rows in order of increasing atomic size. By arranging the elements in this fashion important chemical trends can be recognised for elements appearing in each column. The exact reasons behind this are far too abstruse to be discussed in this simple introduction and are, in any case, incidental to the general subject areas of water and wastewater treatment technology.

Although atoms can be considered indestructible for all purposes other than nuclear power generation, they actually contain many sub-atomic particles, the most important of which can be considered to be *protons*, *neutrons* and *electrons*. The protons and neutrons each weigh 1 unit (Table 1.2) and they form a tiny *nucleus* around which the weightless electrons orbit within specific spatial *shells* or *orbitals*, rather like planets around the sun in a solar system (Figure 1.2).

SUB-ATOMIC PARTICLES
An atom consist of a nucleus composed of protons and neutrons surrounded by electrons.

Table 1.2 – Sub-atomic particles

Particle	Charge	Weight	Location in Atom
Neutron	0	1	Nucleus
Proton	+1	1	Nucleus
Electron	−1	0	Outer shells

Electrons orbit the nucleus in discrete *shells* of different intrinsic energy levels. Each shell (numbered 1, 2, 3 etc.) can be sub-divided into different orbitals (*s*, *p*, *d*, *f*, *g*), as illustrated in Figure 1.2 for oxygen and hydrogen. The way in which electrons fill these orbitals and shells is referred to as the *electon configuration*. Generally, atoms with filled shells are more stable than those whose shells are partially-filled. It is for this reason that the "Group 0" elements in the Periodic Table (Figure 1.1) comprise gases which are extremely unreactive (Helium, Neon, Argon, etc.). These elements, often referred to as the "Inert Gases", all have a filled outer shell.

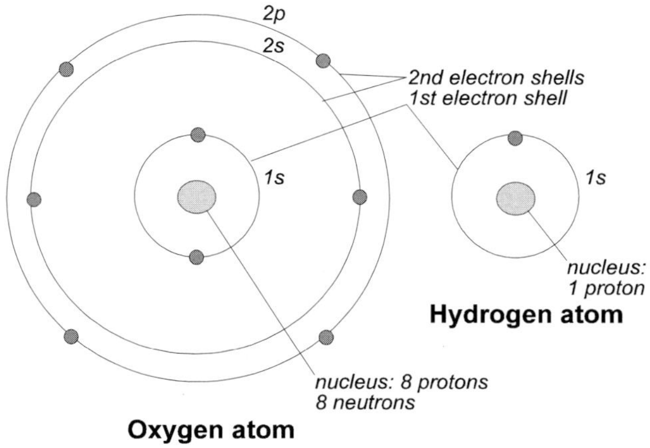

Fig. 1.2 – Atomic structures of oxygen and hydrogen

Protons and electrons carry an electrical charge, the electron being negative and the proton positive. For electroneutrality to be preserved within an atom, the number of electrons in an elemental atom must equal the number of protons. It follows that the simplest of elements need contain no more than one proton and one electron. This is indeed the case for hydrogen. Given that hydrogen contains only one component contributing to its weight, it also follows that the weight of all elemental atoms should be multiples of the weight of a single hydrogen atom. A glance at the third column in Table 1.1 reveals that this is the case for most elements. One important exception is chlorine. It arises because chlorine is actually composed of more than one type of atom which none-the-less all have the same chemical properties; i.e. it has more than one *isotope*. Isotopes are important in radiochemistry, but are incidental to water and wastewater treatment in general.

1.2.2 Molecules and Ions

IONS
Ions are atoms or molecules which have an electric charge as a result of gaining or losing an electron.
Cations are positively charged.
Anions are negatively charged.

In order to become more stable (or less reactive), atoms may either combine with others to form *molecules* or can gain or lose charge (in the form of electrons) to become *ions*. In both cases the net result is usually a change in the number of electrons in the outer shell of the atoms involved. Molecules can form the building bricks of *compounds* in the same way as ions, and some are so stable that for our purposes the group can be considered as indestructible as an atom. An example of a very stable molecule is water (Figure 1.3), formed by what can be an explosive combination of hydrogen and oxygen (Figure 1.2). The nature of the forces which actually bond atoms together in a molecule is quite complicated and, for the purposes of water and wastewater treatment processes, need not be considered further. Suffice to say that these bonds, usually called *valence* bonds, can be broken and reformed to make new compounds by chemical reaction.

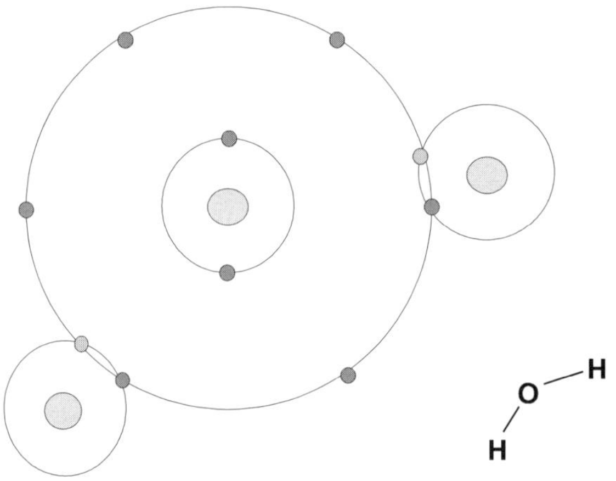

Fig. 1.3 – Molecular structure of water (H_2O)

Figure 1.3 indicates how atoms of hydrogen and oxygen combine to form valence bonds. Note that the sharing of electrons leads to an increase in the number of electrons in the outer shells of each atom, in this case to **8** for the oxygen atom and **2** for the hydrogen atom. The outer shell of each atom is thus filled, and so water is a stable molecule.

ELECTRON TRANSFER
When chemical reactions take place, electrons may be transferred to make the individual constituent atoms more stable. By filling or half filling the outer shell, the atoms become more stable because their electron configuration becomes the same as one of the Inert Gases.

Both atoms and molecules can become *ions* by acquiring a positive or negative electrical charge by respectively losing or gaining electrons. The symbol for an ion is the appropriate chemical symbol plus the number and sign of charges it has acquired superscripted. Positively-charged ions are called *cations* and negatively-charged ions *anions*. For example the sodium ion is a Na atom which has lost an electron to become $\mathbf{Na^+}$, the sodium cation (Figure 1.4), and the sulphate ion is the group SO_4 which has gained two electrons to become $\mathbf{SO_4^{2-}}$, the "sulphate" anion. In both cases, the constituent atoms become stable by having a completely filled outer shell of electrons – either by electron transfer (as with sodium) or both electron sharing and electron transfer (in the case of sulphate). In essence, chemical reactions often arise so that electrons can be transferred to allow the outer electron shell to be filled or half-filled. In Figure 1.4, the transfer of the electron in the outer shell of the sodium to the "space" in the outer shell of the chloride atom leads to greater stability.

The number of charges on an ion is sometimes referred to as the *valency*: hence, the sodium ion has a valency of +1 whilst the sulphate ion has a valency of −2.

ELECTRON CONFIGURATION
When sodium loses an electron the cation formed has the same electron configuration as neon.
When chloride gains an electron the anion formed has the same electron configuration as argon.

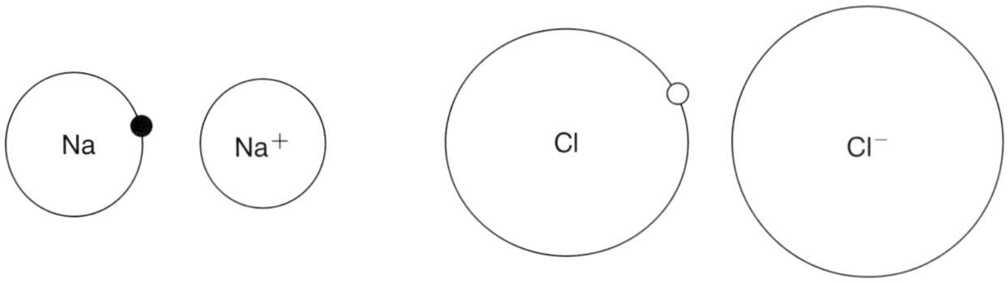

Fig. 1.4 – Ionisation of sodium and chlorine atoms

The common ions are listed in Table 1.3. Note that some ions have names different from the parent element. For example Cl^- is always called "chloride", as opposed to the "chlorine" atom Cl or the chlorine gas molecule Cl_2. Some elements such as iron can form ions with different charges, i.e. "ferric" (Fe^{3+}) and "ferrous" (Fe^{2+}) ions, which have quite different chemical properties from one another and so demonstrate different behaviour in waters and wastewaters.

CHEMICAL NOTATION
No. of atoms in molecule are **subscripted**.
No. and sign of charges on ion are **superscripted**.
NH_4^+ is an ion having a charge of 1+ and containing a nitrogen and 4 hydrogen atoms.
SO_4^{2-} is an ion having a charge of 2− and containing a sulphur and 4 oxygen atoms.

Table 1.3 – Common ions in water treatment

Ion Name	Symbol	Ionic Weight	Equivalent Weight
Aluminium	Al^{3+}	27	9
Ammonium	NH_4^+	18	18
Bicarbonate	HCO_3^-	61	61
Calcium	Ca^{2+}	40	20
Chloride	Cl^-	35.5	35.5
Ferric	Fe^{3+}	56	18.7
Ferrous	Fe^{2+}	56	28
Hydrogen	H^+	1	1
Hydroxyl	OH^-	17	17
Magnesium	Mg^{2+}	24	12
Nitrate	NO_3^-	62	62
Nitrite	NO_2^-	46	46
Sodium	Na^+	23	23
Sulphate	SO_4^{2-}	96	48

1.2.3 Salts

Ions of opposite charge cling together by electrical attraction (or *ionic bonding*) to make compounds called *salts*. Salts which are or can be dissolved to form a solution are sometimes referred to as *electrolytes*. For instance common salt NaCl, is made of Na^+ and Cl^- ions. The total charge of a salt or electrolyte must be zero for electrical neutrality to be preserved. So, for example, two Na^+ ions would combine with a single sulphate ion SO_4^{2-} to form the salt Na_2SO_4 (sodium sulphate). When salts dissolve in water they dissociate into their constituent ions which we write as follows:

SALTS
Salts are compounds made up of cations ionically bonded to anions. When a salt dissolves it dissociates into its constituent cations and anions.

$$\mathbf{NaCl \Leftrightarrow Na^+ + Cl^-} \quad (1.1)$$

$$\mathbf{Na_2SO_4 \Leftrightarrow 2Na^+ + SO_4^{2-}} \quad (1.2)$$

We shall learn more about dissociation in Unit 2. Salts make up most of the material dissolved in natural waters, although certain waters contain significant amounts of dissolved and suspended organic (carbon-based) matter. The nature and chemical reactions of organic matter are extremely complicated and will not be discussed further.

1.2.4 Acids and Bases

ACIDS AND BASES
Acids contain hydrogen as the only cation.
Bases contain hydroxyl as the only anion.
Acids react with bases to produce only salt and water.

A substance which dissociates in water forming hydrogen ions as the only cation is called an *acid*. So hydrochloric acid dissociates as:

$$\mathbf{HCl \Leftrightarrow H^+ + Cl^-} \quad \textbf{(1.3)}$$

And nitric acid:

$$\mathbf{HNO_3 \Leftrightarrow H^+ + NO_3^-} \quad \textbf{(1.4)}$$

A substance which dissociates in water forming hydroxyl ions as the only anion is called a *base* or an *alkali*. The most common are sodium hydroxide (caustic soda) and calcium hydroxide (lime). They dissociate as follows:

$$\mathbf{NaOH \Leftrightarrow Na^+ + OH^-} \quad \textbf{(1.5)}$$

$$\mathbf{Ca(OH)_2 \Leftrightarrow Ca^{2+} + 2OH^-} \quad \textbf{(1.6)}$$

When an acid reacts with an alkali the products of the reaction are a neutral salt plus water:

$$\mathbf{HCl + NaOH \Leftrightarrow NaCl + H_2O} \quad \textbf{(1.7)}$$

1.2.5 The Chemical System of Weights

Ex 1.1
What is the weight of 1 mole of sodium nitrate ($NaNO_3$)?

Each kind of atom has a characteristic weight. Because atoms are built up of units similar to the hydrogen (H) atom, by making the weight of the H atom = 1 unit a system is created whereby each atomic weight is usually a whole number. However, the weight of a single H atom is extremely small, so chemists have calculated how many H atoms weigh 1 gram. The number is called *Avagadro's* number, named after its discoverer, and takes a value of 6.022×10^{23}. This quantity is called a *mole*. The Na atom weighs 23 times as much as the H atom, so that a mole of Na contains the same number of atoms and weighs 23 g, and so on for other elements.

The *molar* or *molecular weight* is the sum of the atoms in the molecule. As electrons are weightless, the ionic weight is the same as the corresponding molecular weight. So, for example, 1 mole of either SO_4 or SO_4^{2-} weighs 96 g, because both species contain one sulphur atom (32 g per mole) and four oxygen atoms (16 g each).

IONIC BALANCE
In any solution the total negative charge of the anions must equal the total positive charge of the cations.

Tables 1.1 and 1.2 include values for the atomic, ionic, and molecular weights. The significance of these will become apparent when considering concentrations of ions in solution, and how they interact. However, when considering solutions of compounds and salts it should always be remembered that, for any solution:

1. The molar weight of each compound is the sum of all the atoms in a mole, and
2. The total charge of the anions must equal the total charge of the cations

1.3 Measurement of Concentration

Section 1.3 requires a study time of about 1 hour

1.3.1 Molar Concentrations

In water treatment we are concerned with a lot of different dissolved impurities in very dilute solution. This calls for specialised units of measurement.

The *mg/l* (milligram per litre) is a unit commonly used in water analysis. As water weighs almost exactly one thousand grams, or one million milligrams per litre, one mg/l is the same as one mg/kg or one *ppm* (part per million). Water chemists use these two systems of concentration measurement fairly interchangeably. Of course, at much higher concentrations when the overall solution density is greater than 1 g/l, the systems are no longer interchangeable as ppm refers to weight per unit weight and mg/l to weight per unit volume.

UNITS
1 litre = 1000 ml
1 ml water weighs 1 g at density of 1 g/l
1 g = 1000 mg
So, 1 ppm = 1 mg per million mg water

It is sometimes more convenient to define the concentration in terms of the number of moles per unit volume, rather than mass. For example, chemists would measure the concentration of a salt in water in moles per litre, which is called *molar* concentration, more usually referred to as the *molarity* of the solution (abbreviation *Mol/l* or simply *M*). So, a 1 M solution of NaCl contains 58.5 g of the salt per litre of solution, whereas the same molar concentration of, say, Na_2SO_4 contains 119 g per litre.

We also have to define what we are weighing. For example, a milliMolar (mMol/l or mM) solution of NaCl contains 58.5 mg NaCl per litre, so a chemical analysis might state, correctly:

NaCl *58.5 mg/l*

Ex 1.2
What is the concentration in g/l as such of a 2M solution of sodium nitrate ($NaNO_3$)?

Since chemists would analyse for Na^+ and Cl^- separately, as NaCl would be expected to be completely dissociated, it is normal and equally correct to say

Na^+ *23 mg/l*

Cl^- *35.5 mg/l*

Written in that form, it is implied that the weight refers to the ion or compound named. Sometimes the analyst spells it out even more clearly by saying:

Na^+ *23 mg/l "as such", or "as Na"*

Ex 1.3
A Solution of calcium sulphate ($CaSO_4$) contains 350 mg/l as SO_4. What is the calcium concentration in mg/l as Ca?

This becomes important when different conventions might be used. Nitrate, ammonia, and other compounds containing the element N (nitrogen) are a common example.

A mMol solution of NO_3^- contains 62 mg/l as NO_3^-, but this could equally be written "14 mg/l as N". Waste waters contain a lot of nitrogenous compounds which change quite easily from one to another, so it is convenient to know the total content of nitrogen, which can only be done by converting all the individual measurements to "mg/l as N" before adding them together. Much the same calculation is carried out with the impurities which cause hardness in a water.

In chemical shorthand square brackets are used to indicate molar concentrations. Thus $[O_2]$ indicates the concentration of oxygen and $[OH^-]$ indicates the concentration of hydroxyl ions, both in moles per litre ($mol.l^{-1}$). A further very common shorthand is used in particular for hydrogen ion concentration:

$$\mathbf{pH = -\log_{10}[H^+]} \qquad \mathbf{(1.8)}$$

"p"
The prefix "p" means "the negative log of an ion concentration has measured in moles per litre". Hence "pH" is the negative log of the molar hydrogen ion concentration.

Ex 1.4
A water sample has a pCa value of 2.3. What is the concentration of calcium in mg/l as such?

Similarly pOH = $-\log_{10}[OH^-]$ and pCa = $-\log_{10}[Ca^{2+}]$. The "p" is short for "power" in the sense of "ten to the power…" In Unit 2 we shall see how pH is used as a measure of the acidity of a solution.

1.3.2 Equivalent Concentration

Ex 1.5
What is the equivalent weight of calcium bicarbonate ($Ca(HCO_3)_2$)?

Many reactions and technologies in water treatment depend on the electrical charges on ions. In order to take account of the ionic charge, we often use *the equivalent weight* (EW), a unit which measures the weight per unit charge. It is calculated by dividing the molar weight (MW) by the *valency* or, in the case of an ionic species, the number of charges on the ion. Thus the equivalent weight is given by:

$$\mathbf{EW = MW/valency} \qquad \textbf{(1.9)}$$

For example, the ion SO_4^{2-} has an ionic weight of 96 g, but as it is *divalent*, i.e. it has a charge of 2, its EW is 48 g. Some ions can carry different numbers of charges, like, for example Fe^{2+} and Fe^{3+}. The EW has to take into account the ionic form under consideration. A complete listing of equivalent weight values is given in Table 2.3.

Concentrations can be expressed as g equiv/l or, more commonly, mg equiv/l which is abbreviated as *meq/l*, (and on the Continent as *m val*). Equal volumes of solutions at the same concentration in meq/l react exactly with one another.

EQUIVALENT
If we work in meq/l or mg/l as $CaCO_3$ then the sum of the cations is always equal to the sum of the anions. This is not the case if we work in mM or mg/l as such.

A similar, rather old-fashioned, unit is still in popular use, the *mg/l as $CaCO_3$*. The *total hardness* and the *alkalinity* (see Unit 2) are often expressed in these terms. It represents the concentration of a substance expressed as if its EW were the same as that of $CaCO_3$. This can cause endless confusion, best avoided by converting to equivalents using the rule:

$$\mathbf{50\ mg/l\ as\ CaCO_3 = 1\ meq/l} \qquad \textbf{(1.10)}$$

Consider a typical water analysis, shown in Table 1.4

Table 1.4 – A typical groundwater analysis

	Concentration			Concentration	
Cation	mg/l	meq/l	Anion	mg/l	meq/l
Ca^{2+}	76	3.8	HCO_3^-	244	4.0
Mg^{2+}	14	1.2	Cl^-	38	1.1
Na^+	35	1.5	SO_4^{2-}	52	1.1
			NO_3^-	21	0.3

The first thing to do with such an analysis is to reduce it to directly useful units. All items shown in the analysis "as such", or whatever, should be converted to meq/l.

$$\mathbf{meq/l = \frac{mg/l \times valency}{MW}} \qquad \textbf{(1.11)}$$

Ex 1.6
A Solution of sodium sulphate (Na_2SO_4) contains 4.7 meq/l. What is the concentration in mg/l as such and in mg/l as Na?

Now concentrations of constituents can be seen in terms of their relative ability to react. Totalling the positively-charged cations should equal the sum of all the negatively-charged anions, for electrical neutrality. This is a good way of checking a water analysis before doing anything with it – if the ions don't balance, there is something wrong. Non-dissociated constituents do not, of course, enter into the ionic balance.

A block diagram (Figure 1.5) is a good way of illustrating how the inorganic constituents of a water relate to one another.

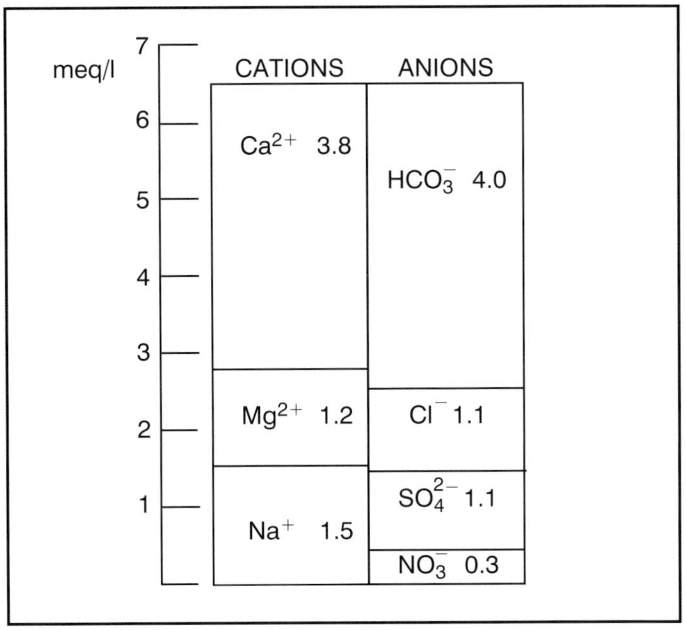

Fig. 1.5 – Block diagram form of data in Table 1.4

Ex 1.7

A laboratory reports the following water analysis all in mg/l as such:

Calcium	*118*
Magnesium	*5*
Bicarbonate	*280*
Chloride	*57*
Sulphate	*96*
Nitrate	*trace*

What is the concentration of sodium in mg/l as such to give a balanced analysis?

1.3.3 Conductivity

H_2O does not conduct electricity, but charged ions in it can move about and carry an electrical current. Very conveniently, the speed with which different mineral ions move in dilute solutions is roughly the same, so measuring the electrical *conductivity* of a water gives a good idea of its total ionic content. The μ*S/cm*, where "S" represents a "Siemen" and equates to a reciprocal ohm ($1S = 1\Omega^{-1}$), is the normal unit used to express conductivity.

A **very crude** approximation relating conductivity to concentration is:

Each meq/l of ions gives a conductivity of 100 μS/cm **(1.12)**

Conductivity is a simple and reliable measurement, and so is another useful check to an analysis of neutral water. H^+ and OH^- ions, which are respectively present in high acidity and high alkalinity waters, move much faster than all other ions in water. The general rule above therefore breaks down progressively outside a range of, say, 6 to 8 pH units.

At higher concentrations than potable water, salts don't dissociate so completely, but empirical relationships can be used to calculate the approximate dissolved salt concentration. Conductivity is such a simple and reliable measurement that it is standard practice for testing the purity of water for such applications as boiler feed, microprocessor wash and pharmaceutical preparation.

Below 0.01 meq/l, the conductivity falls below 1 μS/cm (or 10^{-6} ohm^{-1}), which involves working in decimals. In very pure water, the preferred unit is the resistivity, where:

resistivity = 1/conductivity **(1.13)**

CONDUCTIVITY

Electrical conductivity is a quick and simple measurement that gives an indication of the total dissolved solids (TDS) of water. Typical conductivity levels are:

Sea water	50,000 μS/cm
Potable water	1000 μS/cm
Distilled water	50 μS/cm
Deionised water	1 μS/cm
Ultrapure water	0.06 μS/cm

In very pure water, the conductivity due to the minute concentration of H^+ and OH^- ions becomes significant and has to be taken into account. The resistivity of theoretically pure water at 20°C is **18.6 Megohm cm^{-1}**, due entirely to the H^+ and OH^- ions. At these low ion concentrations conductivity measurement becomes much more difficult, but so do all other methods of analysis and so conductivity remains an essential measurement.

The relationship between conductivity, resistivity and concentration, taken from equations 1.11 and 1.12, is given in Table 1.5

Table 1.5 – Conductivity and concentration

Conductivity (μS.cm^{-1})	Resistivity (MΩ.cm)	Approx Conc'n (meq.l^{-1})
100	0.01	1
10	0.1	0.1
1	1	0.01
0.1	10	0.001

Figure 1.6 shows graphically how the conductivity of solutions of some salts vary with concentration.

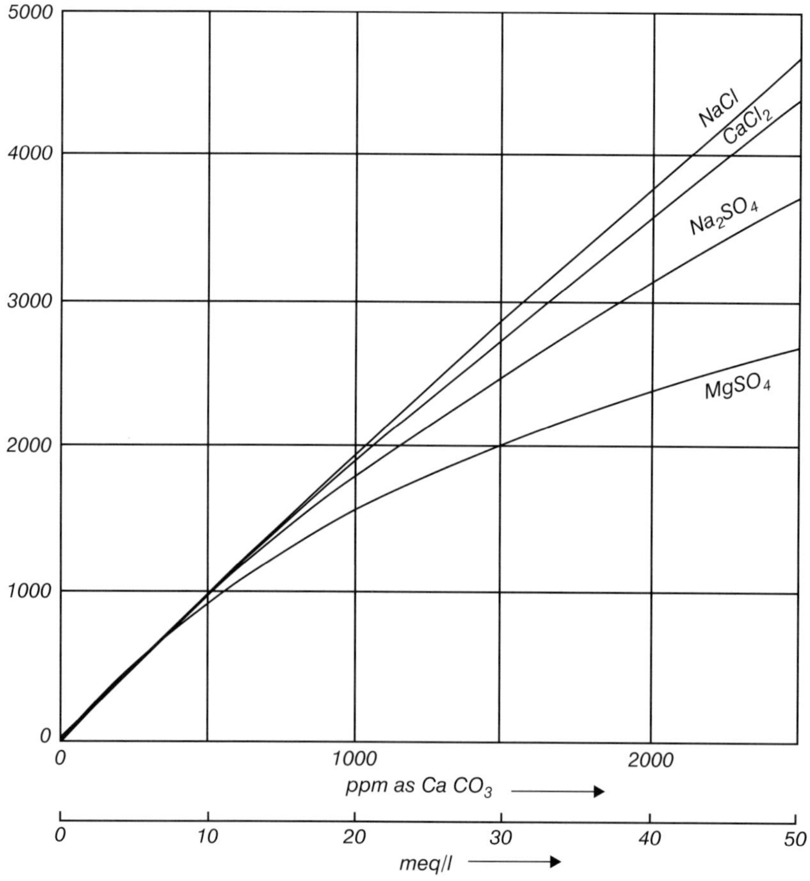

Fig. 1.6 – Conductivity (μS.cm^{-1}) vs concentration

1.3.4 Conversion of Units

Figure 1.7 offers a quick reference for conversion of concentration units between ppm, mM, meq/l, ppm "as $CaCO_3$" and conductivity in μS.cm^{-1}. Note that conversion to conductivity and osmotic pressure involves the use of empirically-derived proportionality coefficients, which have been denoted *y* and *x* respectively.

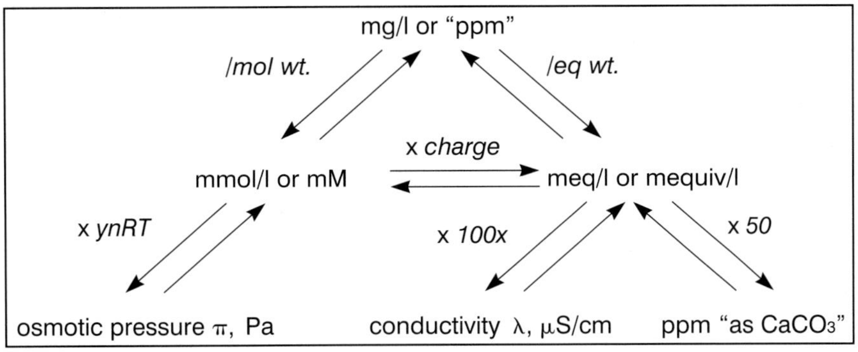

Fig. 1.7 – Conversion of concentration units

Section 1.4 requires a study time of about 2 hours

THE FARADAY
The electrical charge carried by 1 eq is 1 Faraday

1 Faraday = 96493 coulombs/eq

1.4 Electrochemistry

1.4.1 The Faraday

As stated earlier, in any electrolyte solution there must be a charge balance, and a given amount of charge is associated with chemically equivalent amounts of ions. Thus 1 equivalent of sodium ions (23 g) carries the same amount of charge as 1 equivalent of chloride ions (35.5 g). This amount of charge is always known as the Faraday and given the symbol *F*. The value of the Faraday is **96493 coulombs**.

The total charge carried by individual ions is reflected in the current the solution can pass. The total amount of ions (in eq) transported electrically is simply the product of the current and the time passed divided by the Faraday. This is the principal on which Faradaic processes, such as electrodialysis and electrolysis, are based. Given that an amp is defined as a coulomb per second, i.e. the amount of charge passed per unit time, there is direct correlation between current I, time t and the amount of ions passed:

$$\textbf{Moles of ions transferred} = \mathbf{It/nF} \qquad \textbf{(1.14)}$$

where n is the charge on the ion.

1.4.2 Electrode Processes

Many chemical reactions involve the transfer of electrons, the most obvious examples being redox equilibria ("redox" meaning "oxidation + reduction", Unit 1, 1.5.2) such as that between ferrous and ferric iron:

$$\mathbf{Fe^{3+} + \mathit{e} \Leftrightarrow Fe^{2+}} \qquad \textbf{(1.15)}$$

The fact that electron transfer is involved means that, if appropriate reactions are arranged to occur at electrodes, current can flow and electrode potentials measured.

The simplest system to consider is that of a metal electrode in contact with a solution of its ions. For instance a copper plate in a solution of cupric chloride, $Cu/CuCl_2$, gives an equilibrium reaction:

$$\mathbf{Cu^{2+} + 2e \Leftrightarrow Cu} \qquad E^0 = +0.34\ \text{volt at } 25°\text{C} \qquad \textbf{(1.16)}$$

E^0 is the Standard Electrode Potential and the scale is defined such that a hydrogen electrode ($H^+ + e \Leftrightarrow \frac{1}{2}H_2$) has a potential of zero.

It is impossible to measure single electrode potentials: a potential can only be measured with reference to another in the same way that the height of a mountain can only be measured with reference to the sea level. Thus, if two electrodes are connected by a wire and the solutions are connected by a salt bridge, then an electrochemical cell is created and current can flow. For instance, the Cu/Cu^{2+} system above can be combined with a zinc electrode in a solution of zinc chloride, where the equilibrium is:

Ex 1.8
How much dissolved manganese (Mn21) is deposited (as manganese metal) by a current of 1.2 Amps over a 24 hour period?

$$\mathbf{Zn^{2+} + 2e \Leftrightarrow Zn} \qquad E^0 = -0.76\ \text{volt} \qquad \textbf{(1.17)}$$

This gives the so-called Daniell cell (Figure 1.8). The electrode reactions as written above show (from left to right) the reduction of metal ions to give the metal. The standard electrode potential gives a measure of the tendency for the reduction to occur. The more positive E^0, the greater the tendency of the metal ions in solution to acquire electrons and to deposit on the electrode as the metallic element. The values of E^0 given above show that copper ions have a greater tendency to be reduced than zinc ions. So, in the Daniell cell the reactions will proceed such that the metallic zinc of the electrode is oxidised to give zinc ions in solution and copper ions are reduced and deposit as metallic copper at the copper electrode. In order to sustain this process, electrons must flow in the external circuit and ions must be transferred through the salt bridge between the electrode compartments.

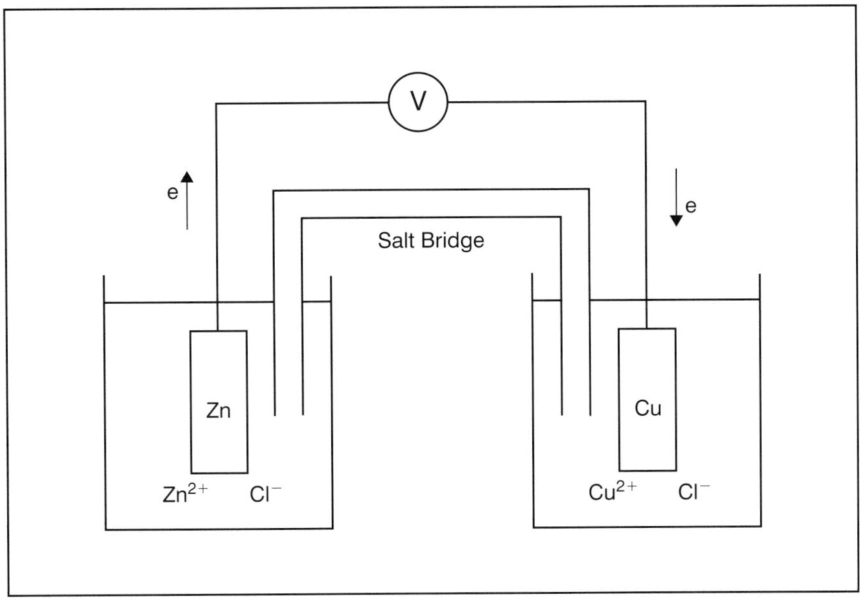

Fig. 1.8 – The Daniell cell

The two electrode reactions can be combined to give the overall equilibrium reaction:

			E^0	
Zn	⇔	**$Zn^{2+} + 2e^-$**	**+0.76 V**	**(1.18)**
$2e^- + Cu^{2+}$	⇔	**Cu**	**+0.34 V**	**(1.19)**
$Zn + Cu^{2+}$	⇔	**$Zn^{2+} + Cu$**	**+1.10 V**	**(1.20)**

Note that the zinc electrode reaction is written in the opposite direction, so that the sign of E^0 is reversed from that given in Equation 1.17

The fact that the overall reaction, as written, has a positive potential, means that the formation of zinc ions should be favoured. In the Daniell cell electrons in the external circuit should flow from the zinc anode (where oxidation occurs) to the copper cathode (where reduction occurs). It would be possible to apply an external potential to oppose this current and, in the case of the Daniell cell, a value of 1.10 V would be needed to prevent the flow of electrons altogether and this is the standard cell potential. In principle, once one electrode potential is known, then any other can be found by assembling the appropriate cell and measuring the overall cell potential. As mentioned before, all electrode potentials are, by convention, expressed relative to the standard hydrogen electrode.

NERNST EQUATION

The Nernst Equation predicts the voltage generated by a cell as a function of the equilibrium constant

$$E = E^0 - \frac{RT}{nF} \ln K$$

Ex 1.9

A Daniell cell operating at 25°C has an initial copper sulphate concentration of 4 M. What voltage will it generate when the zinc concentration has increased to 1 M?

Because zinc has a greater tendency to dissolve than most common metals (more positive E^0), it is often used to protect other metals against corrosion – either as a sacrificial anode or by coating (galvanizing) the metal with a layer of zinc.

The potential, or voltage generated by a cell is given by the *Nernst equation*:

$$E = E^0 - \frac{RT}{nF} \ln K \qquad \textbf{(1.21)}$$

where R = the universal gas constant (8.3 J.°K^{-1}.mol^{-1})
T = temperature, °K (where °K = °C + 273)
K = equilibrium constant (Unit 2, 3.1): ratio of product to reactant species.

The equilibrium constant is a function of the concentration of the reactants and products. If these are at equilibrium, $E = 0$ reflecting that, at equilibrium, the system is at a minimum free energy and so cannot perform any useful work.

NB Sections 1.4.4–1.4.6 are at ***advanced*** *level.*

1.4.3 Redox Potentials

As well as the electrodes discussed above, there are several other types known. If an inert metal such as platinum or gold is placed in a solution containing oxidised and reduced species, then an electrode potential is established, for which a Nernst-type expression applies.

The potential measured by such an electrode is often known as an oxidation-reduction potential *(ORP)* or simply Redox potential. It is sometimes given the symbol E_H. This concept is quite useful in natural waters since it enables the possibility of certain reactions to be established and whether, under given conditions, substances will exist in their oxidised or reduced states.

An important example is the case of iron in water containing dissolved oxygen. Oxygen is a powerful oxidising agent and its reduction can be written as the following half-cell reaction:

$$O_2(aq) + 4H^+ + 4e^- \Leftrightarrow 2H_2O \qquad E^0 = 1.27\,V \qquad \textbf{(1.22)}$$

Combining this with the oxidation of ferric iron gives:

$$O_2(aq) + 4Fe^{2+} + 4H^+ \Leftrightarrow 4Fe^{3+} + 2H_2O \qquad E^0 = 0.50\,V \qquad \textbf{(1.23)}$$

Using Equation 1.21 we can calculate the ratio of ferric to ferrous iron for a known oxygen concentration and pH value. For a dissolved oxygen content of 5 mg/l or 1.6×10^{-4} M (somewhat below the saturation value), the result becomes:

$$E = 0.50 - 0.059\left(\log\frac{[Fe^{3+}]}{[Fe^{2+}]} + pH + 0.95\right)\text{volts} \qquad \textbf{(1.24)}$$

When E = 0 the system is at equilibrium and, for pH 2, ferric iron should be present at more than 100,000 times the concentration of the ferrous form. At higher pH values the ratio becomes lower, but there are then complications due to hydrolysis. Such calculations can only give concentrations of species at equilibrium and can give no information on rates of reactions.

1.4.4 Ion Selective Electrodes

The response of electrodes to ions in solution gives possibilities of determining ion concentrations using electrochemical techniques. The best-known example is the *glass electrode*, which is very widely used to measure pH. This essentially consists of a thin glass membrane separating the test solution from a solution of HCl in contact with a silver/silver chloride electrode (see below). The action of pH electrodes depends essentially on the ion exchange properties of glass: a surface layer becomes hydrated and hydrogen ions can replace sodium ions in this hydrated layer. It can be shown that the response of such an electrode is given by a Nernst-type expression:

$$E = E^0 + \frac{RT}{F} \ln[H^+] \qquad (1.25)$$

That is

$$E = E^0 - 59.2 \times \text{pH mV} \qquad (1.26)$$

For each unit increase of pH value, the potential of the glass electrode becomes more negative by about 59 mV.

In common with all specific ion electrodes, the glass electrode responds strictly to activity rather than concentration of the appropriate ion. Only in fairly dilute salt solutions can the concentration be derived directly from the electrode response.

By using modified forms of glass it is possible to prepare electrodes which respond selectively to cations such as sodium and potassium. Again, Nernstian responses are found.

The silver/silver chloride electrode consists of a silver wire with a coating of AgCl. In aqueous solutions, this electrode responds to chloride ion activity and it can be shown that the response is given by:

$$E = E^0 - \frac{RT}{nF} \ln[Cl^-] \qquad (1.27)$$

This electrode finds wide application, not only for monitoring chloride levels, but also as a component of many electrode systems.

By combination of ion exchange materials (liquids or membranes) with various electrode types, it is possible to manufacture devices which respond selectively to a wide range of ions. Commercially-available electrodes include those for nitrate, fluoride, lead and many others. These electrodes are never entirely *specific* for a given ion – there is always the problem of interference by other ions, although this may be negligible in some cases.

1.4.5 Amperometric Electrodes

The electrode methods mentioned above are potentiometric, since they involve the measurements of electric potentials. Another set of techniques relies on current measurement to determine concentrations of various species in water. These make use of electrode reactions where the electrode is polarised. In this condition the reactant concentrations at the electrode is depleted since transport of reactant to the electrode is the rate-determining step. The rate of the reaction is governed by the rate at which the reactant can be transported to the electrode, which, under standard conditions, is proportional to the concentration of the reactant. Because an electron transfer process is involved, the current through the cell will also be proportional to the reactant concentration and so can be used as an analytical tool. Appropriate cells can either be electrolytic, where an external voltage is applied, or galvanic, where the potential is generated by making the polarised electrode part of a galvanic cell.

In water and wastewater analysis, a very common example of such a technique is the dissolved oxygen electrode. In one version, a lead anode and a silver cathode are employed. At the anode, metallic lead dissolves to give Pb^{2+}, releasing electrons which pass through the circuit to the silver cathode and drive the reduction of oxygen to hydroxide. Some of the OH^- ions migrate to the anode and react with Pb^{2+} to give $Pb(OH)_2$. The overall reaction is:

$$\mathbf{O_2 + 2Pb + 2H_2O \Leftrightarrow 2Pb(OH)_2} \qquad \textbf{(1.28)}$$

The rate of this reaction is controlled by the rate of supply of oxygen to the silver anode, which occurs across an oxygen-permeable membrane. This rate is proportional to the dissolved oxygen concentration in the bulk solution (provided that this is well stirred to avoid depletion of oxygen outside the membrane). Hence the current generated gives a measure of the oxygen concentration.

The determination of chlorine in water is often carried out by an amperometric titration, which is based on similar principles.

1.5 Chemical Calculations in Water Treatment

Section 1.5 requires a study time of about 2 hours

1.5.1 Stoichiometry

In Unit 2 reaction kinetics and equilibria and how they influence the way in which chemicals react with one another are considered. For the moment it is assumed that chemical reactions proceed rapidly to completion, such that all the reactants are completely converted to products. Whilst something of a simplification, this assumption is very often used in order to calculate the amounts of reagents needed and, if appropriate, by-products generated during chemical treatment. Such a calculation demands a knowledge of the chemical reactions involved; since all chemical reactions proceed on a molecular level the relative amounts of chemicals involved can be calculated from the ratio of the number of molecules of each chemical species involved in the governing chemical equation. This forms the basis of stoichiometry, which defines the ratio to which chemicals react based on the weight of the individual molecules involved in the reaction. Put simply, the amount of reagent A required to react completely with constituent B can be calculated if the number of molecules of A that react with B is known. For example, for the precipitation reaction:

$$\mathbf{Al^{3+} + 3OH^- \Leftrightarrow Al(OH)_3\downarrow} \qquad \mathbf{(1.29)}$$

the stoichiometric ratio of hydroxide ions (OH^-) to aluminium ions (Al^{3+}) is 3 : 1, implying that dosing of water with aluminium salts has a profound effect on the solution pH because hydroxide ions are used up, tending to make the solution more acidic (Unit 2, 2.4.3).

1.5.2 Balanced Equations

One of the fundamental principles of chemistry is that the products and reactants of a chemical equation should be in perfect balance: i.e. there should be no net gain in either material or electrical charge. The latter arises from the *redox* principle, which states that oxidation of one species of a chemical reaction must necessarily give rise to reduction of another. For example, the oxidation of Fe^{2+} can be expressed in many forms – all of which obey the fundamental rule of material and charge balance:

$$\mathbf{Fe^{2+} - 1e^- \Leftrightarrow Fe^{3+}} \qquad \mathbf{(1.30)}$$

$$\mathbf{4Fe^{2+} + O_2 + 10H_2O \Leftrightarrow 4Fe(OH)_3\downarrow + 8H^+} \qquad \mathbf{(1.31)}$$

$$\mathbf{2Fe^{2+} + \tfrac{1}{2}O_2 + 4OH^- + H_2O \Leftrightarrow 2Fe(OH)_3\downarrow} \qquad \mathbf{(1.32)}$$

In the first example (Equation 1.30) the reaction is expressed simply as a transfer of an electron from the ferrous ion Fe^{2+} to an unidentified source. The second and third equations identify the source as the oxygen molecule, which is then reduced to the hydroxide ion (OH^-) and subsequently combines with the ferric ion (Fe^{3+}) to form ferric hydroxide ($Fe(OH)_3$), sometimes referred top as ferric floc. Whilst Equations 1.31 and 1.32 may look different, they are both valid expressions of the same redox reaction.

The rules governing stoichiometry are not always straightforward. However, it is convenient to regard all elements in compounds as having an *oxidation state* or *oxidation number* (O.N.), which can be the charge on the ion or the valency of the covalently-bonded element. For example, iron can have an O.N. of zero (in its elemental form), +2 (in its ferrous form) or +3 (in its ferric form). Carbon almost always has an O.N. of +4 in the organic and inorganic compounds it forms. Chlorine usually takes an O.N. of −1, for example in hydrochloric acid (HCl) and tetrachloroethane (CCl_4), but can also assume an O.N. of +1, for example in hypochlorite (OCl^-), to +7, for example in perchloric acid ($HClO_4$). Finally, nitrogen can assume oxidation states of −3 and +5 in its most stable forms (in ammoniacal substances and nitrate respectively) but can also assume many other oxidation states in more reactive substances, such as nitrous oxide (N_2O, O.N. = +1), nitric oxide (NO, O.N. = +2), nitrite (NO_2^-, O.N. = +3) and nitrogen dioxide (NO_2, O.N. = +4).

Oxidation number is often indicated as a Roman numeral superscript as in Fe^{III} or Cr^{VI}.

BALANCING EQUATIONS
Equations are balanced in 3 steps:
OXIDATION NUMBER
CHARGE BALANCE
MATERIAL BALANCE

1.5.3 Methodology for Balancing Equations

For most chemical reactions encountered in water treatment the following six basic rules can usually be applied:

1. There can be no net change in the quantity of each element.
2. There can be no net gain or loss of charge. Thus, in redox reactions: (a) the sum of the ionic charges on one side of the equation must be the same as that on the other, and (b) the sum of the total oxidation numbers on either side of the equation must also be the same.
3. Oxygen takes a value of −2 unless it is either (a) in the elemental state as O_2 or O_3, in which case it is zero, or (b) in the *peroxo* state (for example in hydrogen peroxide H_2O_2) in which case it is −1.
4. Hydrogen take a value of +1 unless it is either (a) in the elemental state as H_2, in which case it is zero, or (b) in the hydride form (H^-), in which case it is −1.
5. In their most stable forms:
   Halogens (F, Cl, Br and I) take an O.N. of −1 or, if covalently bonded to a nitrogen or oxygen atom, a value of +1,
   Nitrogen takes values between −5 (as in cyanide CN^-) to +5 (as in nitrate NO_3^-),
   Alkaline or *Group I* metals (Li, Na, K, Rb and Cs) take an O.N. value of +1,
   Alkaline Earths or *Group II metals* (Mg, Ca, Sr, Ba) take an O.N. value of +2.
6. If there is an imbalance of material, it can usually be assumed that it can be made up with water or its products H^+ or OH^-.

Ex 1.10

Dissolved Fe^{2+} reacts with permanganate (MnO_4^-) to produce the solids ferric floc ($Fe(OH)_3$) and MnO_2 – the most stable form of manganese. Thus:

Reactants: $Fe^{2+} + MnO_4^-$
Products: $Fe(OH)_3 + MnO_2$

Write a balanced equation for this reaction

Balancing of any chemical equation proceeds in three stages:

1. **Oxidation number calculation**
   O.N. values normally calculated from the ion charge and/or the assumption of values of +1 and −2 for combined hydrogen and oxygen respectively.
2. **Charge balance**
   Ratio of the reactants and products from the redox principle
3. **Material balance**
   Quantity of additional atoms (O and H) required for equation to balance

1.6 Key Reactions in Water Treatment

Section 1.6 requires a study time of about 2 hours

This section comprises some examples of chemical reactions as applied to water and wastewater treatment, to demonstrate the validity of the simple stoichiometric approach to designing chemical processes. These reactions will be covered in more detail in Module 4.

1.6.1 pH Adjustment and Neutralisation

One of the most important chemical processes in water and wastewater treatment is the adjustment of pH. Calculation of dosing rates demands a basic understanding of logarithms since, according to Equation 1.8:

$$\mathbf{pH = -log_{10}[H^+]}$$

However, the Law of Mass Action and equilibrium thermodynamics (Unit 2, 2.3.1 & 2.4.2–2.4.3) states that:

$$\mathbf{pH + pOH = 14} \qquad \mathbf{(1.33)}$$

This means that an acid solution contains only low levels of hydroxide and vice versa. Equation 1.33 also determines that the minimum concentration of hydroxide and acid takes place under conditions of:

$$\mathbf{pH = pOH = 7; [H^+] = [OH^-] = 10^{-7}} \qquad \mathbf{(1.34)}$$

Hence, neutralisation of a solution, i.e. adding the same molar concentration of acid as there is base or vice versa, produces a solution containing 10^{-7} M each of H^+ and OH^- (or 0.0001 ppm and 0.0017 ppm of each respectively). Clearly, these levels are very low but not zero.

The logarithmic relationship has significant implications. The addition of 10^{-6} moles per litre of acid or hydroxide to a neutral solution changes the pH by one unit. If the same dose is added to a solution at pH 4 the pH barely changes, since the acid concentration at pH 4 is 1000 times that at pH 7 and a 100 times greater than the added dose.

1.6.2 Iron Precipitation

Many borehole waters contain quantities of iron dissolved as Fe^{2+}. As already indicated, Fe^{2+} can be oxidised to produce a Fe(III) precipitate using an oxidant such as air (see Equations 5.2–5.4). Other more powerful oxidants, such as permanganate (MnO_4^-), can also be used to achieve this conversion more rapidly:

$$\mathbf{3Fe^{2+} + MnO_4^- + 5OH^- + 2H_2O \Leftrightarrow 3Fe(OH)_3\downarrow + MnO_2\downarrow} \qquad \mathbf{(1.35)}$$

According to the above equation, 1 mole of permanganate (molecular weight 119 g/mol MnO_4^-) can oxidise 3 moles of Fe^{2+} (56 g/mol). The reaction is base catalysed so if at moderate to high alkalinities the reaction can be assumed to go to completion, then, according to Equation 1.33, 119/3 ppm of permanganate is required to oxidise 56 ppm of iron, and hence 0.71 ppm per ppm of iron. This, of course, assumes that there are no side reactions which might deplete the permanganate. In practice, permanganate is rarely used for this duty, partly because of its tendency to produce a pink colouration in the product water.

1.6.3 Carbonate Precipitation

Treatment of raw water with lime ($Ca(OH)_2$) is sometimes used to soften water. When lime is added to a solution having high levels of hardness (Ca^{2+} and Mg^{2+} ions and alkalinity (HCO_3^- ions), precipitation of calcium carbonate ($CaCO_3$) takes place according to the following chemical reaction:

$$\mathbf{Ca(OH)_2 + Ca(HCO_3)_2 \Leftrightarrow 2CaCO_3\downarrow + 2H_2O} \quad (1.36)$$

According to this reaction, 1 mole of calcium bicarbonate (molecular weight 162 g) reacts with 1 mole of lime (molecular weight 74 g) to produce 2 moles of calcium carbonate product (molecular weight 100 g). If the reaction can be assumed to go to completion, this would imply that 0.46 mg/l of lime would need to be added for each 1 mg/l (1 ppm) of calcium bicarbonate for all the calcium to be precipitated out of solution. This is rarely the case in practice, when an excess of reagent is normally required. The addition of 0.55 mg/l of lime in the above case would equate to a *20% stoichiometric excess*.

1.6.4 Ion Exchange

Ion exchange is a sorption process in which ions in the solution are actually exchanged with ions associated with an insoluble synthetic resin. An example is the well-established softening process for removing hardness from water:

$$\mathbf{2\,(\mathit{r}\text{-}Na) + Ca^{2+} \Leftrightarrow (\mathit{r})_2Ca + 2Na^+} \quad (1.37)$$

where r represents the resin matrix. The above process allows hardness, in the form of calcium and magnesium ions, to be removed from the water to be replaced by the less troublesome sodium ion. However, the process can be reversed at much higher concentrations, behaviour which derives from equilibrium thermodynamics, allowing the resin to be regenerated using brine solutions. In general, exchange will always take place if the ions in solution are of (a) higher charge and (b) greater weight than those associated with the resin provided the solution is of low ionic strength, the preference being governed by the selectivity coefficient. This is actually another example of the Law of Mass Action and equilibrium thermodynamics.

In order to calculate the quantity of ion exchange material required between regenerations under a given set of conditions the following information is required:

- the level of hardness (calcium and magnesium) in the raw water
- the capacity of the ion exchange material for exchanging hardness.

These are both normally expressed in equivalents per unit volume (for dissolved hardness) or per unit weight (for the resin capacity), so that the two measurements are comparable on a charge for charge basis. The amount of resin required is then simply given by the total amount of hardness per run divided by the resin capacity:

$$\mathbf{W\,(kg) = \frac{run\ time\ (h) \times flow\ rate\ (l/h) \times hardness(eq/l)}{capacity\ (eq/kg)}} \quad (1.38)$$

The amount of regenerant can also be calculated along similar principles. The stoichiometric excess of regenerant required to remove the adsorbed hardness in order for the resin to attain its original capacity is given by the *regeneration ratio*:

$$R = \frac{\text{amount of regenerant required to regenerate resin (eq)}}{\text{amount of adsorbed hardness on resin (eq)}} \qquad (1.39)$$

R is always greater than unity, but all ion exchange processes aim to reduce the value of R as much as possible in order to limit chemical usage and the amount of effluent produced.

1.6.5 Chlorination of Ammoniacal Solutions

The reaction between chlorine and ammonia is not fully understood. Equations proposed to describe the chlorination and oxidation of ammonia include:

$$NH_3 + HOCl \Leftrightarrow NH_2Cl + H_2O \qquad (1.40)$$

$$NH_3 + 2HOCl \Leftrightarrow NHCl_2 + 2H_2O \qquad (1.41)$$

$$2NH_3 + 3HOCl \Leftrightarrow N_2 + 3H^+ + 3Cl^- + 3H_2O \qquad (1.42)$$

$$NH_4^+ + 4HOCl \Leftrightarrow NO_3^- + 6H^+ + 4Cl^- + H_2O \qquad (1.43)$$

The stoichiometric ratio of chlorine to ammonia can thus vary between 1 : 1 and 4 : 1, according to the above expressions, depending how far the reaction proceeds.

The stoichiometry of the oxidation reactions, Equations 1.42 and 1.43, which describe a process which is usually referred to as breakpoint chlorination, is pertinent to both potable and waste water treatment because of its implications with respect to both the nature of the end products and the chlorine consumption. Whilst it is desirable to produce more innocuous nitrogen product (N_2) generated from reaction 1.42, it is not usually possibly to exercise the necessary solution chemical control required to only partially oxidise the ammonia. The formation of significant quantities of the chlorine-substituted ammoniacal compounds NH_2Cl and $NHCl_2$, collectively called chloramines, is not always desirable. This issue, and others surrounding the controversial subject of disinfection by-product formation, will be discussed in more detail in a later module.

1.6.6 Biochemical Oxidation

The significance of reaction stoichiometry is not limited solely to chemical reactions. In the activated sludge process microorganisms can convert organic matter to oxidation products and, ultimately, energy according to the following governing equations:

$$COHNS + O_2 + \text{nutrients} \Leftrightarrow CO_2 + NH_3 + C_5H_7NO_2 + \text{others} \qquad (1.44)$$

$$C_5H_7NO_2 + 5O_2 \Leftrightarrow 5CO_2 + 2H_2O + NH_3 + \text{energy} \qquad (1.45)$$

where the "COHNS" in Equation 1.44 represents organic matter containing carbon, oxygen, hydrogen, nitrogen and sulphur, and "$C_5H_7NO_2$" represents new bacterial cells. Hence, according to Equation 1.44, 5 moles of oxygen are needed to allow bacteria to aerobically convert 1 mole of cell material to carbon dioxide, water and ammonia. This equates to a weight ratio of 160 : 113, or 1.42 : 1, oxygen : cells for complete oxidation. This then directly relates to the biochemical oxygen demand (BOD) of the wastewater, which is simply 1.42 times the concentration by weight of the cellular material.

1.7 Self Assessment Questions

SAQ 1.1 A laboratory reports the following water analysis all in mg/l as such unless indicated otherwise:

Calcium	76
Magnesium	14
Sodium	35
Bicarbonate	200 mg/l "as $CaCO_3$"
Chloride	38
Nitrate	31

What is the concentration of sulphate in mg/l as such to give a balanced analysis?

SAQ 1.2 The reaction between dissolved calcium chloride and sodium sulphate is known to produce a precipitate of calcium sulphate. What mass of precipitate is produced if a stoichiometric excess of sulphate is added to a 200 $mg.l^{-1}$ "as $CaCO_3$" solution of calcium?

SAQ 1.3 The ammonium ion (NH_4^+) can be fully oxidised to nitrate (NO_3^-) by hypochlorous acid (HOCl), which is reduced to chloride. Write a balanced equation for this reaction, and then calculate the concentration in mg/l of hypochlorite (OCl^-) required to fully oxidise 15 ppm "as N" of ammonia.

SAQ 1.4 Aluminium chloride reacts with sodium hydroxide to produce aluminium hydroxide solid as one of the products. What mass of sodium hydroxide is required to react with 2kg of aluminium chloride and what mass of aluminium hydroxide is precipitated?

SAQ 1.5 If 10 ppm of aluminium chloride is added to water at pH 5 and completely reacts to form aluminium hydroxide, what is the new pH?

[NB: The aluminium takes the hydroxide from the water molecule, liberating H^+ in the process]

SAQ 1.6 What current is required to deposit 15 g of iron over a 48 hour period from a ferrous nitrate solution?

SAQ 1.7 By how much will the actual electrode potential change for a ten-fold increase in the Fe^{2+}/Fe^{3+} concentration ratio?

1.8 Glossary of Terms

Acid	Compound capable of dissolving to release H^+ ions.
Alkali	Compound capable of dissolving to *neutralise* H^+ ions.
Anion	Negatively-charged *ion*.
Atom	Fundamental particle of an element, containing protons, neutrons and electrons.
Atomic weight	Weight, in grams, of one mole of an element. The *relative* atomic weight of an element is its atomic weight divided by the atomic weight of hydrogen (1 g).
Avagadro's number	Number of atoms in a mole = 6.022×10^{23}.
Cation	Positively-charged ion.
Compound	Substance containing a combination of elements.
Covalent bond	Joining of atoms through sharing of electrons in their outer shell. Associated largely with organic molecules.
Electrolyte	Compound capable of dissociating in water to form ions.
Electron	Fundamental negative electrical charge, with zero weight, existing in outer extremities of an atom in discrete shells.
Electron configuration	The way in which electrons are arranged in orbitals and shells around the nucleus, which then determines the number of electrons in the outer shell.
Element	Simple substance of matter. All matter can be resolved into elements, of which there are 109 in total.
Equivalent (eq)	Unit of ionic charge concentration: eq/l = mole/l $\times$ valency.
Formula	Shorthand notation used to express the components of any compound. The number of each component in the formula is subscripted, e.g.: Na_2SO_4 contains $2 \times Na$ (sodium), and $4 \times O$ (oxygen), $Ca(NO_3)_2$ contains $2 \times NO_3$ (nitrate), and thus $6 \times O$.
Inorganic	Compounds, often ionic and therefore dissociating in water, which do not contain oxidisable carbon.
Ion	An atom or molecule which has gained or lost one or more electrons to become charged. The ion charge is superscripted, e.g.: Na^+ (sodium cation) has a charge/valency of +1. SO_4^{2-} (sulphate anion) has a charge/valency of −2.
Ionic bonding	Electrostatic interaction between ions, formed by transfer of one or more electrons between the atoms of a salt.
Isotopes	Forms of an element having a differing atomic structures but displaying identical chemical behaviour.
Mole	1 mole = Avagadro's number $\times$ 1 atom.
Molar weight	Weight of one mole of a substance.
Molarity	Concentration in moles/l.
Molecule	Group of atoms connected by ionic or covalent bonding.
Neutralisation	Reduction in alkaline or acid conditions.
Neutron	Fundamental neutrally-charged subatomic particle, having a relative mass of 1. Found in the nucleus of an atom.
Nucleus	Core of atom containing neutrons and protons.
Orbital	Component of shell.
Organic	(Oxidisable) compound of carbon. These compounds usually contain hydrogen, nitrogen and/or oxygen.
Periodic Table	Table (Figure 1.1) containing all elements in order of their atomic weight. These weights increase along each row of the Table, and the elements in each column (or *group*) have similar chemical properties because they contain the same number of electrons in the outer electron shell.

Proton	Fundamental positively-charged subatomic particle, having a relative mass of 1. Found in the nucleus of an atom. It is also the alternative name given to the H^+ ion, as an H^+ ion contains only 1 proton.
Salt	Compound containing a combination of ions.
Shell	Layer of electron(s) around nucleus of atom. These contain individual discrete orbitals, each of which can hold a pair of electrons well filled.
Valence bond	Bond existing between atoms in molecules.
Valency	Number of positive or negative electrical charges on an ion.

1.9 Solutions to Exercises

Ex 1.1 Atomic weights from Table 1.1: Na 23, N 14, O 16.
Molecular weight = 23 + 14 + (3 × 16) = 85

Ex 1.2 Molecular weight of $NaNO_3$ = 85 (Ex 1).
2 M = 2 mol/l = 2 × 85 = 170 g/l.

Ex 1.3 MW SO_4^{2-} = 96 (Table 1.1)
350 mg/l as SO_4^{2-} = 350/96 mmol/l = 3.65
$CaSO_4$ has 1 mol Ca^{2+} and 1 mol SO_4^{2-}
3.65 mmol/l SO_4^{2-} = 3.65 mmol/l $CaSO_4$ = 3.65 mmol/l Ca^{2+}
MW Ca^{2+} = 40 (Table 2.1)
3.65 mmol/l Ca^{2+} = 3.65 × 40 = 146 mg/l as Ca

Ex 1.4 pCa = 2.3 = $-\log_{10}[Ca^{2+}]$
$[Ca^{2+}] = 10^{-2.3} = 5.01 \times 10^{-3}$ mol/l = 5.01 mmol/l
MW (Table 1) Ca = 40
$[Ca^{2+}]$ = 40 × 5.01 = 200 mg/l as Ca

Ex 1.5 MW (Table 1.1) Ca 40, H 1, C 12, O 16
MW $Ca(HCO_3)_2$ = 1 × 40 + 2 × (1 + 12 + (3 × 16)) = 162
Valency = 2 so EW = 162/2 = 81

Ex 1.6 MW (Table 1.1) Na 23, S 32, O 16
MW Na_2SO_4 = (2 × 23) + 32 + (16 × 4) = 142
Valency = 2 so EW = 142/2 = 71
4.7 meq/l = 4.7 × 71 = 333.7 mg/l as such
EW Na = 23/1 = 23
4.7 meq/l = 4.7 × 23 = 108.1 mg/l as Na

Ex 1.7 EW (Table 1.3) Ca 20, Mg 12, Na 23,
HCO_3 61, Cl 35.5, SO_4 48
Convert to meq/l
Ca = 118/20 = 5.9
Mg = 5/12 = 0.4
HCO_3 = 280/61 = 4.6
Cl = 57/35.5 = 1.6
SO_4 = 96/48 = 2.0
Total anions = 4.6 + 1.6 + 2.0 = 8.2
Total cations = 5.9 + 0.4 + Na = 8.2
Na = 1.9 meq/l
Na = 1.9 × 23 = 44 mg/l as such

Ex 1.8 Mass (g) = {27.5(g/eq)/96,500(eq/C)} × 1.2(C/s) × time (s)
If t = 24 hours (86,400 s), then
mass = 27.5 × 1.2 × 86,400/96,500 = 29.5 g

Ex 1.9 Values to substitute in Equation 1.21:
E^0 = 1.10 V and n = 2 (Equation 1.19)
R = 8.314, F = 96,493
T = 25 + 273 = 298°K
1 mol Cu^{2+} generates 1 mol Zn^{2+}
original $[Cu^{2+}]$ = 4 so if $[Zn^{2+}]$ = 1, $[Cu^{2+}]$ = 4 − 1 = 3
E = 1.10 – (8.314 × 298/(2 × 96,493)) × ln(1/3)
E = 1.10 – 0.013 × ln(1/3) = 1.11 V

Ex 1.10 *Oxidation number calculation*

Assuming appropriate O.N. values for H and O:

Fe: From +2 in reactant to +3 in product, a change of +1

Mn: From +7 in reactant to +4 in product, a change of −3

Hence for redox balance, **1 mole of Mn reacts with 3 moles of Fe**

Charge balance

Reactants: $(3 \times +2) + (1 \times -1) = +5$

Products: zero

Hence for charge balance, add **5 moles of negative charge to reactants** (assumed to be hydroxide ion) *or* **5 moles positive charge to product** (as H^+).

Material balance

So far, we have:

$$\mathbf{3Fe^{2+} + MnO_4^- + 5OH^- \Leftrightarrow 3Fe(OH)_3\downarrow + MnO_2\downarrow}$$

Obviously Fe and Mn both balance. The oxygen and hydrogen count for the above equation is:

O: 2 mole deficit on the reactant side

H: 4 mole deficit on the reactant side

Hence the final balanced equation is:

$$\mathbf{3Fe^{2+} + MnO_4^- + 5OH^- + 2H_2O \Leftrightarrow 3Fe(OH)_3\downarrow + MnO_2\downarrow}$$

2. Chemical Kinetics and Equilibria

Contents

Aims and Objectives

After studying these notes you should be able to:

1. explain in your own words the following terms and concepts:

   The law of Mass Action
   Henry's Law
   alkalinity
   hardness
   pH
   equilibrium constant
   solubility product
   Langelier Saturation Index

2. carry out calculations for predicting:

   pH
   solubility and precipitation of salts
   pH adjustment by acid or alkali dosing
   gas solubility
   scaling tendency

After studying these notes it is not essential for you to:

1. memorise any of the constants
2. memorise the equations

It is important that you are able to correctly complete all the self assessment questions at the end of these notes.

Essential Prerequisites

Prior to undertaking this Unit it is important that you have completed and understood Unit 1, Fundamentals of Water Chemistry including atoms, molecules, ions, acids and bases and concentration measurement in mg/l, mol/l, meq/l and mg/l as $CaCO_3$.

2.1 Introduction

Chemistry is traditionally divided into three areas: *Organic*, *Inorganic* and *Physical*. The first of these deals with reactions of molecular compounds containing carbon, including solvents and polymers. Inorganic chemistry considers the chemistry of salts, including both metals and non-metals. In both these branches of chemistry it is actual changes in the compounds that are of importance; for example the addition of caustic soda to a solution of iron sulphate to produce iron hydroxide (ferric floc) or the production of ion exchange materials by sulphonation of a 3D polymer with sulphuric acid.

Physical chemistry, on the other hand, does not focus on any one type of chemical, but on the way in which all chemicals behave under a given set of conditions. Specifically, physical chemistry considers (a) the relative levels of chemicals in solution once they have undergone a chemical reaction, and (b) how quickly that reaction proceeds. These two key branches of physical chemistry are respectively referred to as *equilibria* and *kinetics*.

In the following discussion, very simple but important examples of chemical equilibrium will be discussed in turn. All these equilibria are governed by one simple universal law: the *Law of Mass Action*.

Section 2.2 requires a study time of about 1 hour

2.2 Kinetics

2.2.1 Reaction Rates

Chemical reactions do not happen instantaneously, indeed some reactions, like oxidation of iron in damp atmospheres, are very slow. It is important to know exact rates of a governing reaction in order to calculate the correct size for a reactor. Many reactions take place in a sequence of individual reaction steps all of which proceed at different rates. The slowest reaction is the one which effectively controls the overall rate of reaction and this is the *rate determining step*.

2.2.2 Reaction Order

LAW OF MASS ACTION
The Law of Mass Action tells us that, the rate of chemical reaction is proportional to the concentrations of the reactants as defined by Equation 2.2.

The *Law of Mass Action* tells us that the rate of a chemical reaction is proportional to the concentrations of the reactants. For the general reaction:

$$aA + bB \Rightarrow cC + dD \quad (2.1)$$

the rate of change of concentration of reactants and products can be represented as d(concentration)/dt. Considering all concentrations as molarities:

$$r = \frac{1}{d}\frac{d[D]}{dt} = \frac{1}{c}\frac{d[C]}{dt} = -\frac{1}{a}\frac{d[A]}{dt} = -\frac{1}{b}\frac{d[B]}{dt} \quad (2.2)$$

where r is the rate of reaction. The rate at which the reaction proceeds is a function of the initial concentrations of the individual components:

$$r = k[A]^m[B]^n \quad (2.3)$$

where k is the *rate constant* for the reaction
m is the *order* of the reaction with respect to component A
n is the *order* of the reaction with respect to component B

DIFFERENTIAL CHANGES
The change in something is prefixed "d".

If this change is with respect to time it is denoted "d/dt".

So, change in concentration c with respect to time is "dc/dt".

The (negative) change in the molar concentration of reactant "A" with respect to time is: "−d[A]/dt".

The rate constant is characteristic of the particular reaction and depends on operating conditions such as temperature, pressure etc. For most reactions m and n will have values of 0, 1 or 2. If m = 1 and n = 0 the reaction is said to be zero order with respect to B and first order with respect to A. If both m and n = 1 the reaction is first order with respect to both and A and B *and second order overall*. Note that if both m and n = 0 then the rate of reaction is independent of the initial concentrations of the components. Not all reaction kinetics are as simple as this – in some cases m and n are non-integer exponents and in some the exponents are algebraic expressions.

2.2.3 Integrated Rate Laws

DIFFERENTIATION & INTEGRATION
Differentiation (with respect to time) yields the rate at which the component in the system is changing at a specific point.

Integration between two points in time yields the total change in the component over that period of time.

Equation 2.3 defines the rate of reaction but if we wish to use it to predict the concentration of a component at a given time we have to integrate the equation. If we consider the rate at which one of the reactants, say A, is consumed we can write:

$$-\frac{d[A]}{dt} = k[A]^m \quad (2.4)$$

For a zero order reaction (m = 0) integrating equation 2.4 gives:

$$\Delta[A] = [A]_0 - [A]_t = kt \quad (2.5)$$

where k is in $mol.l^{-1}.s^{-1}$. We define $t_{1/2}$ as the *half life*, that is the time taken for the initial concentration $[A]_0$ to be reduced to $0.5[A]_0$. Thus:

$$\mathbf{t_{1/2} = [A]_0/2k} \qquad (2.6)$$

So if the reaction rate is independent of the reactant concentration then decay is linear with time and the half life is directly proportional to the initial concentration.

If the reaction is first order (m = 1) then integrating Equation 2.4 gives:

$$\mathbf{\ln([A]_t/[A]_0) = -kt} \qquad (2.7)$$

that is

$$\mathbf{[A]_t = [A]_0 e^{-kt}} \qquad (2.8)$$

where k is in s^{-1}. So if the reaction rate is proportional to the reactant concentration then decay is exponential with time. The product concentration, though rarely of interest in water and wastewater treatment, is then given by:

$$\mathbf{[C] = [A]_0(1 - e^{-kt})} \qquad (2.9)$$

and the half life given by:

$$\mathbf{t_{1/2} = \ln 2/k = 0.693/k} \qquad (2.10)$$

So, in this case, the half life depends only on the rate constant and not on concentration.

Ex 2.1
What is the rate constant of a first order reaction in which the concentration decays by 25% in 5 minutes?

Finally, if the reaction is second order (m = 2) then integrating Equation 2.4 gives:

$$\mathbf{1/[A]_t - 1/[A]_0 = kt} \qquad (2.11)$$

that is

$$\mathbf{[A]_t = [A]_0/(1 + kt[A]_0)} \qquad (2.12)$$

where k is in $l.mol^{-1}.s^{-1}$ and the half life is now given by:

$$\mathbf{t_{1/2} = 1/k[A]_0} \qquad (2.13)$$

so that a second order reaction is characterised by an increasing half life for the decaying reactant over the course of the reaction.

Section 2.3 requires a study time of about 1 hour

2.3 Equilibrium

2.3.1 Reversible Reactions

Unless the products of chemical reaction are removed, the reaction can normally proceed in either direction. Consider a general *reversible* chemical reaction:

$$\mathbf{aA + bB \Rightarrow cC + dD} \qquad \textbf{(2.14)}$$

Here A and B are reactants, C and D are products and a, b, c and d represent the number of moles of each component existing at equilibrium. The Law of Mass Action states that the rate of the forward reaction is given by:

$$\mathbf{r_f = k_f[A]^a[B]^b} \qquad \textbf{(2.15)}$$

and the rate of the backward reaction by:

$$\mathbf{r_b = k_b[C]^c[D]^d} \qquad \textbf{(2.16)}$$

In these equations the square brackets [ ] denote the concentration of a component in moles per litre. Strictly, the function should not be defined in terms of the concentration but of the *activity* of the ion, a_i, which is defined as:

$$\mathbf{a_i = \gamma_I \times [ion]} \qquad \textbf{(2.17)}$$

where γ is the *activity coefficient*. The activity coefficient is, itself, a function of the total ionic concentration. In dilute solutions – the sort of concentrations normally found in water – the activity coefficient is very close to unity and thus molar concentration units are applicable.

Eventually the reaction will reach *equilibrium*, a stable state in which it appears as if no more product is being formed. In fact, reactants and products are continually being formed at equilibrium at equal rates, such that the reaction appears to be at rest. This means that the rates of the forward and backward reactions (Equations 2.15 and 2.16) are equal so that the equilibrium condition is defined by:

$$\mathbf{K = \frac{[C]^c[D]^d}{[A]^a[B]^b}} \qquad \textbf{(2.18)}$$

K, which is the ratio of the two rate constants, k_b/k_f, is known as the *equilibrium constant*.

CHEMICAL EQUILIBRIUM
The Law of Chemical Equilibrium tells us that, at equilibrium, the rates of the forward and backward reactions are equal and defines the Equilibrium Constant (Equation 2.18).

For any chemical reaction at equilibrium at a given temperature and pressure, K is constant. So, if the reaction conditions were to be disturbed by, say, adding more B or taking away some of C, the concentration of components would readjust to produce the same value of K.

The *Law of Chemical Equilibrium* expressed in Equation 2.18 allows the calculation of the concentration of any one component of a chemical system, provided the value of the equilibrium constant at that temperature and pressure is known. This has important implications in the removal impurities from water.

2.3.2 Weak Acids and Bases

We have so far considered so called *strong electrolytes* which, in solution, dissociate completely into their constituent ions. Some electrolytes, common ones include carbonic acid, acetic acid and ammonium hydroxide, do not dissociate completely and are called *weak electrolytes*. Dissociation can be regarded as a reversible, equilibrium reaction, with a *dissociation constant*, K_a. For the dissociation of acetic acid:

$$CH_3COOH \Leftrightarrow CH_3COO^- + H^+ \quad (2.19)$$

K_a is defined as:

$$K_a = \frac{[H^+][CH_3COO^-]}{[CH_3COOH]} \quad (2.20)$$

Because dissociation constants of weak acids are usually very small, it is convenient to express them in the logarithmic form (as in Unit 1) that is $pK_a = -\log_{10} K_a$. Tabulated values are normally given in this form and the stronger the acid the *lower* is the pK_a value. Equation 2.20 implies that when the pH value equals pK_a, the concentrations of HA and A^- are equal. In other words, the dissociation is half complete. At lower pH values (more acid solutions) the dissociation is reduced and at higher values it is increased. When the pH is greater than pK_a by more than two units, the dissociation is greater than 99% complete.

For weakly basic substances, such as ammonia, acquisition of a hydrogen ion from water can occur:

$$NH_3 + H_2O \Leftrightarrow NH_4OH \Leftrightarrow NH_4^+ + OH^- \quad (2.21)$$

Here we define K_b as:

$$K_b = \frac{[NH_4^+][OH^-]}{[NH_3]} \quad (2.22)$$

The concentration of water can be taken as constant and does not need to be included in this expression. The protonated ion, NH_4^+ in this example, is known as the *conjugate acid* and can be regarded as undergoing a dissociation equilibrium:

$$NH_4^+ \Leftrightarrow NH_3 + H^+ \quad (2.23)$$

with a corresponding acid dissociation constant K_a (of around 5×10^{-10}) defined as in Equation 2.20 for acetic acid.

Ex 2.2
How much free ammonia (NH_3) is there in water at pH3?

Section 2.4 requires a study time of about 2 hours

2.4 Solution

2.4.1 Dissolution of Salts

Salts are composed of molecules which are made up of positive and negative ions held together by electrical attraction. When a salt dissolves in water the ions become free to move independently. This *dissociation* is represented thus for sodium chloride:

$$\mathbf{NaCl \Leftrightarrow Na^+ + Cl^-} \qquad \mathbf{(2.24)}$$

DISSOLUTION OF SALTS
When salts dissolve in water they dissociate completely into their constituent ions. The resulting positive and negative charges will always be equal.

Some ionic compounds dissociate very readily and, in the same way as acids, are thus described as strong electrolytes, or strongly dissociated. This means that in dilute solution (such as in natural waters) it may be assumed that no molecules remain because each has been completely separated into discrete cations and anions. Further, because the salts are electrically neutral (that is the number of positive electrical charges on the cations equals the number of negative electrical charges on the anions), the solution will also be electrically neutral.

If a mole of NaCl and a mole of KNO_3 are dissolved in 100 litres of water, a solution containing only Na^+, K^+, Cl^- and NO_3^- ions is obtained. This solution is identical to one made in 100 litres of water with one mole each of $NaNO_3$ and KCl. Similarly, a solution containing 0.2 moles per litre of NaCl and 0.1 moles per litre of $MgSO_4$ is identical to one containing 0.1 mole each of $MgCl_2$ and Na_2SO_4. This was originally discussed in Unit 1.

Other ions, however, are weakly dissociated and the degree to which the parent molecule is split depends on the circumstances. Examples which are particularly important in natural waters are the salts of bicarbonate (HCO_3^-), and silicate ($HSiO_3^-$), whose behaviour will be described in a subsequent Unit.

SOLUBILITY PRODUCT
The solubility of a salt can be calculated from its *solubility product*:
$K_{SP} = [C]^m[A]^n$

The solubility of most solids *increases* with temperature, although there are some important exceptions to this. At the limit of its equilibrium solubility, a compound is said to form a saturated solution in the water. If the ions reach a concentration higher than saturation, they precipitate. This can happen if a saturated solution is cooled.

The dissolution of a solid salt is represented by the equation:

$$\mathbf{C_mA_n \Leftrightarrow mC^{n+} + nA^{m-}} \qquad \mathbf{(2.25)}$$

where C is the cation and A the anion and the equilibrium constant is given by:

$$\mathbf{K_{SP} = [C]^m[A]^n} \qquad \mathbf{(2.26)}$$

or

$$\mathbf{\log K_{SP} = m \log[C] + n \log[A]}$$

Ex 2.3
A water containing 120 mg/l Ca^{2+} (Mol Wt 40) is saturated with fluoride. What is the concentration in mg/l of fluoride (Mol Wt 19) if the solubility product of CaF_2 is 3.2×10^{-11}?

This is yet another example of the Law of Mass Action, only this time it is the term for the concentration of solid salt $[C_mA_n]$ that is ignored because it is always zero. The constant K_{SP} is called the *molar solubility product* of the salt. If the product of the concentrations of the ions dissolved in water exceed the solubility product then precipitation will occur. For example adding sodium sulphate to a solution of calcium chloride is likely to cause the solubility product of calcium sulphate to be exceeded and calcium sulphate to precipitate.

When precipitation occurs molecules of the salt settle on some solid surface – either the containing wall, or on particles which form nuclei. If there aren't enough nuclei, the solution may become *supersaturated*, and then any slight change in conditions can promote precipitation. The occurrence of super-saturation is an important example of a kinetic effect in water.

Values for the equilibrium solubility of some salts at ambient temperatures are given in Table 2.1. Note that these refer to pure water, as the presence of other contaminants can influence the solubility of a compound. The table includes values of solubility products.

Table 2.1 – Some solubilities of salts at 20°C

Salt	Solubility mg/l	Solubility product K_{SP}
NaCl	263,000	1900
$NaHCO_3$	65,000	9.3
$CaSO_4$	2980	1.9×10^{-4}
$CaCO_3$	1.2	8.7×10^{-9}

2.4.2 The Dissociation of Water

As already explained, salts, acids and bases dissociate into their constituent ions when they dissolve in water and it is assumed that, at normal concentrations, the water molecule is, to all intents and purposes, undissociated. However, in pure water the H_2O molecule itself dissociates giving measurable numbers of H^+ and OH^- ions.

$$\mathbf{H_2O \Leftrightarrow H^+ + OH^-} \tag{2.27}$$

H^+ ions, sometimes referred to as *protons*, constitute the "acidity" of a solution, whereas OH^- (or *hydroxide* ions) contribute to, but do not make up entirely, the *alkalinity* of the solution. In fact, it will be seen later that other anions are generally more important in determining the alkalinity because the alkalinity is defined as the total concentration of anions that react with acid.

DISSOCIATION OF WATER
The water molecule dissociates into hydrogen and hydroxide ions under the Law of Mass Action. The equilibrium constant, K_w for the dissociation is 10^{-14}, thus:

$$pH + pOH = 14$$

Whenever an H^+ and an OH^- ion collide they re-unite to H_2O. The rate at which such collisions occur depends on the numbers in solution and a kinetic equilibrium is set up when it equals the rate at which the water is dissociating. As stated earlier, at kinetic equilibrium two equal and opposite reactions are going on simultaneously, so it appears as if nothing is happening. As water splits very slowly, equilibrium is set up at a very low concentration of H^+ and OH^- ions. If the water is completely pure, then equal numbers of H^+ and OH^- ions are produced, so that the water is neither acid nor alkaline, but neutral. In neutral water, there are only 10^{-7} moles per litre of each ion, i.e. pH 7.

The equilibrium shown in Equation 2.27 is influenced by the presence of acids or alkalis. Acids are actually simply salts containing a proton or more instead of the normal cation(s). When these protons are released (by dissociation) into water they make the solution acidic. A *strong* acid, such as hydrochloric (HCl) or sulphuric (H_2SO_4) will release virtually all its protons into the water whereas a *weak* acid, such as orthophosphoric (H_3PO_4) or acetic (CH_3COOH), will release far fewer. Similarly, the addition of a strong alkali, such as caustic soda (NaOH), will also have an effect.

Suppose we added enough HCl to pure water to supply ten additional H^+ ions for every one already there, then the probability of H^+ and OH^- ions colliding increases tenfold. So does the rate at which they re-unite to form H_2O, but as there is no new source of OH^- ions the concentration of OH^- ions falls. The rate of collisions, which will equal the rate at which water splits, is restored roughly when the number of OH^- ions is reduced to one-tenth its previous value. The ionic (and electrical) balance of the solution is made up by the Cl^- ions. With more H^+ ions than OH^- ions in solution the water becomes acidic. Conversely, the addition of a strong alkali, such as NaOH, increases the concentration of OH^- thereby reducing that of H^+ and creating an alkaline solution.

This logic can be expressed arithmetically: in order to equal the rate of water splitting, the result of multiplying the concentrations of H^+ and OH^- together must always come to the same number. This number is sometimes called the equilibrium constant for water, K_w, or the *ionic product of water*, and remains at 10^{-14} regardless of the acidity or alkalinity of the water. Hence:

$$[H^+] \times [OH^-] = K_w = 10^{-14}\,mol^2.l^{-2} \qquad \textbf{(2.28)}$$

It can be seen that, if the concentration of water itself, $[H_2O]$, is simply ignored because it remains constant (a valid assumption at low concentrations), then Equation 2.28 is simply an expression of the Law of Chemical Equilibrium (Equation 2.18).

2.4.3 pH

pH
The pH of a solution is defined as:

$$pH = -\log_{10}[H^+]$$

Neutral solutions have $pH = 7.0$
$pH < 7$ indicates an increasingly acidic solution
$pH > 7$ indicates an increasingly alkaline solution.

As discussed in Unit 1, pH is a measure of the acidity, that is a measure of the concentration of hydrogen ions. "pH" is a shorthand for "the negative logarithm of the hydrogen ion concentration in moles per litre". Hence, if the concentration of H^+ ions is 10^{-7} mol/l, then the pH is 7. This system is convenient for describing low concentrations and can be used for other ions – for example traces of sodium in high-pressure boiler feed are measured as pNa. It is also sometimes convenient to use pK for the negative log of the equilibrium constant.

According to Equation 2.28, the product of the concentrations of H^+ and OH^- ions is 10^{-14} at all times. It follows that:

$$pH + pOH = 14 \qquad \textbf{(2.29)}$$

So the pH value also tells us what the pOH is: if pH is 4, then the pOH must be $14 - 4 = 10$.

Ex 2.4
A dilute solution of sodium hydroxide (Mol Wt 40) has a pH value of 9.5. What is the concentration in mg/l of sodium hydroxide? What would be the new pH if 1 mg/l of hydrochloric acid (HCl) were to be added? [HINT: the H^+ reacts with the OH^- according to the reverse of Equation 2.27)

At room temperature, pH 7 is neutral. The dissociation of water is faster at higher temperatures, which means that in hot water there will be more H^+ and OH^- ions. The pH will then be lower but the water will still be neutral because pOH will also be lower.

2.4.4 Dissolution of Gases

Molecules in a gas move around freely in space, like manic ping-pong balls. They exert pressure by hitting the walls. In a mixture each gas exerts its own partial pressure independently of the others, and the total gas pressure is the sum of all the partial pressures. As air is composed mostly of O_2 and N_2 in the ratio of 1:4, in dry air at 1 bar the total pressure is due to O_2 at 0.2 bar, plus N_2 at 0.8 bar, total 1 bar.

The partial pressure is of course a measure of the rate with which its molecules bombard the walls and therefore the rate at which they fall back into a water surface. That in turn dictates the equilibrium amount which dissolves in water, because of the constant exchange of these gases between the two phases:

$$N_2(air) \Leftrightarrow N_2(water)$$

$$O_2(air) \Leftrightarrow O_2(water)$$

The solubility of gases in water is therefore directly proportional to its partial pressure in the atmosphere. This relationship is another example of the Law of Mass Action, and is normally expressed in the form of *Henry's Law*:

$$p_i = Hx_i \qquad (2.30)$$

where x_i is the mole fraction of gas dissolved in water (i.e. moles gas per moles water), and p_i is its partial pressure in the atmosphere. **H** is *Henry's Constant* for that gas at that temperature, and its units depend on those taken for *p* and x (Table 2.2). However, it is normally expressed in atmospheres (column 2 in Table 2.2). Henry's constant **H** rises, and gases become less soluble, with increasing temperature, but even cold water dissolves very little O_2.

HENRY'S LAW CONSTANT
The units of H, the **Henry's Law constant**, vary according to those chosen for the partial pressure p_i and liquid concentration c_i of the component i partitioned between the two phases.

Table 2.2 – Henry's constant at 20°C

	H atm	H atm.l/mg	H atm.l/mmol
Oxygen	4.3×10^4	0.024	0.77
Air	7.71×10^4	0.046	1.39
Carbon dioxide	151	6.17×10^{-5}	2.72×10^{-3}
Chloroform	170	2.55×10^{-5}	3.06×10^{-3}
Bromoform	35	2.4×10^{-6}	6.29×10^{-4}
Ozone	5×10^3	1.87×10^{-3}	0.090
Chlorine	767	1.95×10^{-4}	0.014

Ex 2.5
What is the partial pressure of chloroform in the atmosphere adjacent to water contaminated with 12 μg/l of chloroform?

Dalton's Law tells us that the partial pressure of a gas is simply the total pressure, P, multiplied by the concentration of the gas in the gas phase, that is:

$$p_i = y_i P \qquad (2.31)$$

Here y_i is the gas phase mole fraction of gas i. Substituting in Equation 2.30 gives a relationship for the gas and liquid phase concentrations:

$$x_i = y_i(P/H) \qquad (2.32)$$

Once water molecules get into a vapour they behave just like any other gas molecule. The higher the temperature, the more water molecules will have made it into the air, and the fewer gas molecules into the water. At ambient temperature there isn't much water in air, but with increasing temperature there will be more and more, exerting an increasing partial pressure. As the temperature approaches boiling point, the partial pressure of water in the vapour begins to dominate. At the boiling point the partial pressure of the water alone equals the total pressure, and the partial pressure of the other gases becomes zero. Then gases dissolved in water can only go one way – out of the water into the vapour, where the steam which is generated sweeps them away and the water becomes totally degassed.

GAS EQUILIBRIUM
At equilibrium the gas and liquid phase concentrations of a volatile species are related by:
$x_i = y_i\,P/H$

The decreasing solubility (i.e. the highest concentration the water is able to dissolve) of gases with temperature provides a means of ridding water of gaseous (or highly volatile) contaminants. An alternative and more widely employed method is decreasing the partial pressure of the contaminant in the gas phase

Ex 2.6
Water is saturated with air at 4 atm pressure. The pressure is then reduced to 1 atm, resulting in the release of air bubbles. What quantity of air (in mg/l) is released under these conditions?

by creating a vacuum, thereby forcing dissolved gases out of solution in much the same way as increasing the temperature. Gas, and in particular oxygen, solubility also plays a key part in governing aerobic biological processes. These processes rely on the transfer of oxygen from the gas to the liquid phase in order to allow biological degradation reactions to take place. Henry's Law is therefore important in determining the rate of mass transfer of gases in unit processes which rely on the introduction and/or the removal of gases in water.

2.5 Alkalinity and Hardness

Section 2.5 requires a study time of about 2 hours

2.5.1 Bicarbonate and Carbonate

The compounds CO_2, HCO_3^- and CO_3^{2-} are actually linked by a complex series of reversible reactions leading to kinetic equilibria. The reactions are normally collectively referred to as carbonate chemistry. They begin with CO_2 in the atmosphere. When atmospheric air comes into contact with pure water, CO_2 dissolves in water, the amount varying with the partial pressure according to Henry's Law.

$$\mathbf{CO_2(gas) \Leftrightarrow CO_2\ (dissolved)} \qquad (2.33)$$

Dissolved CO_2 combines with water to form carbonic acid (H_2CO_3) in another reversible reaction:

$$\mathbf{CO_2(dissolved) + H_2O \Leftrightarrow H_2CO_3} \qquad (2.34)$$

Carbonic acid is an extremely unstable compound, so unstable that it only exists temporarily before breaking down to form carbon dioxide and water, or else dissociating to form acid and bicarbonate. It none-the-less exists in solution, if only fleetingly, and forms part of the sequence of reactions of carbonate, all of which are subject to the Law of Mass Action when the reactions are all in equilibrium.

CARBONIC ACID
Natural waters always contain dissolved CO_2 which forms carbonic acid, H_2CO_3. This dissociates to form bicarbonate ions HCO_3^-. HCO_3^- is usually the largest anionic component in the analysis and Ca^{2+} the main cationic component.

Carbonic acid dissociates in two successive reversible stages:

$$\mathbf{H_2CO_3 \underset{\textit{1st Dissociation}}{\Leftrightarrow} H^+ + HCO_3^- \underset{\textit{2nd Dissociation}}{\Leftrightarrow} 2H^+ + CO_3^{2-}} \qquad (2.35)$$

Accordingly the Law of Mass Action states that the equilibrium constants for the three successive reactions are given by:

$$K_{\text{dissolution}} = \frac{[H_2CO_3]}{[CO_2][H_2O]} \qquad (2.36)$$

$$K_{\text{1st dissociation}} = \frac{[H^+][HCO_3^-]}{[H_2CO_3]} \qquad (2.37)$$

$$K_{\text{2nd dissociation}} = \frac{[H^+][CO_3^{2-}]}{[H^+][HCO_3^-]} = \frac{[H^+][CO_3^{2-}]}{[HCO_3^-]} \qquad (2.38)$$

In Equation 2.36 the concentration of the water can be ignored, for identical reasons as those discussed in Section 2.4.2. On the other hand, the instability of H_2CO_3 is such that its concentration is always very low, such that, for practical purposes, it can be ignored. This implies that the equilibrium constant for the dissolution reaction is close to zero, whereas the value for the first dissociation is almost infinitely large. Multiplying the first two constants together produces another constant:

$$K_{\text{dissolution+1st dissociation}} = \frac{[H^+][HCO_3^-]}{[CO_2]} \qquad (2.39)$$

which is easier to measure directly because all the components are stable.

Values for all the constants are given in Table 2.3, and results generated from these values shown in Figure 2.1. The figure demonstrates how strongly dependent on pH the concentration of all carbonate-derived constituents is, in much

Ex 2.7
Ultrapure water is stored under atmospheric air which contains 0.03% CO_2. What is the equilibrium pH? What would be the new pH if 2 mg/l of sodium hydroxide were to be added?

[HINT: You'll need to refer to Table 2.3 and Equation 2.39]

the same way that of chlorine species is. Also, the physical nature of these species (i.e. gas, dissolved or solid) has important repercussions.

Table 2.3 – Carbonate equilibrium constants

Equation	Reaction	Value at 20°C
5.1	Dissolution of CO_2	2.72 atm.l/mol
5.7	Dissolution + 1st dissociation	4.14×10^{-7}
5.6	Dissociation of bicarbonate	3.96×10^{-11}

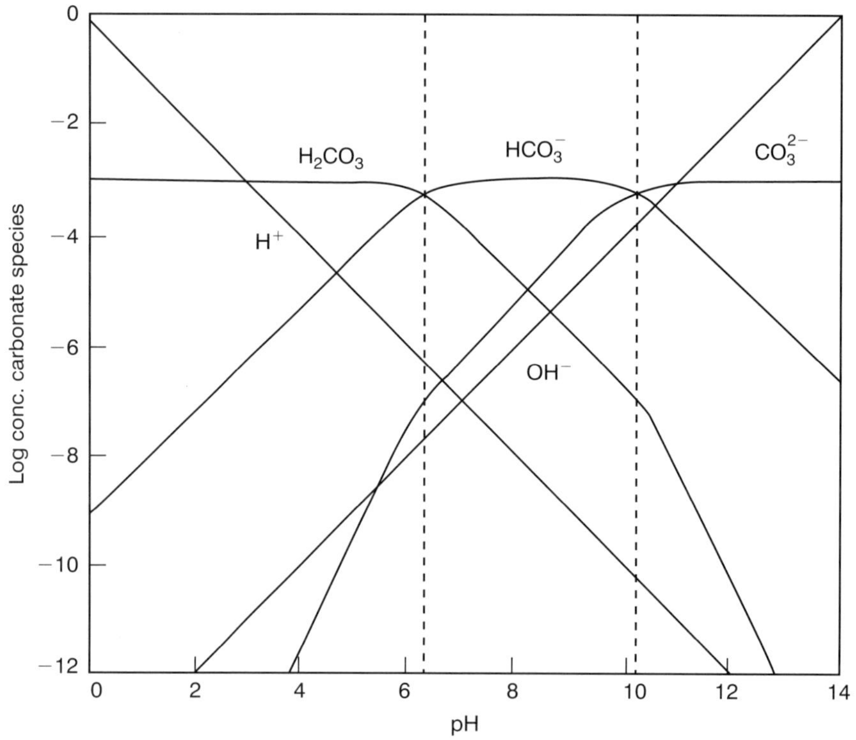

Fig. 2.1 – Concentration distribution for carbonate species

Suppose atmospheric CO_2 is dissolving in pure water. The dissolved CO_2 forms H_2CO_3, which can be ignored. It may be assumed that CO_2 splits directly to H^+ and HCO_3^- ions in the first dissociation stage of Equation 2.35. In pure water, the concentration of H^+ and HCO_3^- ions at which equilibrium is reached is a function of the partial pressure (or concentration) of CO_2. This is clearly demonstrated by Equation 2.39. There thus exists a small concentration of HCO_3^- which can dissociate further, according to the second dissociation stage of Equation 2.35. However, at this pH, the reverse reaction to this second dissociation is so fast that, at these concentrations, equilibrium is all towards the left. The level of CO_3^{2-} is so low that it cannot be detected. Once again, this is obvious from the Law of Mass Action, as given in Equation 2.39.

If the water is then dosed with alkali, the OH^- ions introduced take out H^+ ions by combining with them to form H_2O, so raising the pH. This reduces the availability of H^+ ions, and as a result it reduces the rate at which both first and second dissociation stages go from right to left. As the pH rises, that is, as the H^+ concentration falls, therefore, the concentration of HCO_3^- rises in order to restore equilibrium.

2.5.2 Alkalinity

Bicarbonate, carbonate and hydroxide ions all react with mineral acids and are, therefore, "alkaline salts". The term *alkalinity*, when used in water treatment parlance, refers to the total concentration of these salts and does not mean that the pH is necessarily above 7 (though it usually is).

Classically, alkalinity is determined by titration of a water sample against standard sulphuric acid solution using coloured indicators. The two indicators used are methyl orange, which changes colour at a pH of about 4.5 (corresponding roughly to the first dissociation of carbonic acid) and phenolphthalein, which changes colour at pH 8.2 (the second dissociation of carbonic acid). As we have seen, above pH 8.2 and below pH 4.5 there is no bicarbonate present. This means that the alkalinity measured by titration with phenolphthalein (*"P Alkalinity"*) measures hydroxide (rarely present in natural waters but common in chemically treated waters and wastewaters) plus carbonate. The titration to the methyl orange end point (*"M Alkalinity" or "Total Alkalinity"*) measures hydroxide plus carbonate plus bicarbonate. As a rule of thumb we can assume that half of the carbonate is titrated to pH 8.2 and half between 8.2 and 4.5 – examination of Figure 2.1 shows that this is not too inaccurate – and this allows us to calculate the relative concentrations of the carbonate species from the two measurements, P and M. Table 2.4 shows how.

ALKALINITY
The total alkalinity (*M Alkalinity*) of a water is the sum of all the salts which react with acid, that is HCO_3^-, CO_3^{2-} and OH^-.
P Alkalinity is the sum of the OH^- plus half the HCO_3^-.

Ex 2.8
A water analysis reports a water with P alkalinity 120 mg/l as $CaCO_3$ and M alkalinity 290 mg/l as $CaCO_3$. What is the carbonate concentration?

Table 2.4 – Relationship between P and M alkalinity

P Alk	OH^-	CO_3^{2-}	HCO_3^-
nil	nil	nil	M
<½M	nil	2P	M – 2P
½M	nil	M	nil
>½M	2P – M	2(M – P)	nil
M	M	nil	nil

Equation 2.39 can be rewritten in logarithmic form as:

$$\log K = \log[H^+] + \log[HCO_3^-] - \log[CO_2] \quad (2.40)$$

Using the definition $pH = -\log[H^+]$ and substituting the value of K from Table 2.3:

$$\log_{10}\frac{[HCO_3^-]}{[CO_2]} = pH - 6.38 \quad (2.41)$$

In the case of normal waters in which Ca^{2+} is the major cation Equation 2.41 holds quite well between pH 4.5 and 8.2 and can be used to predict changes in pH resulting from acid or alkali dosing. Adding one mole of acid converts one mole of bicarbonate alkalinity to one mole of carbon dioxide:

$$HCO_3^- + H^+ = CO_2 + H_2O \quad (2.42)$$

so that dosing x mole of acid gives:

$$\log_{10}\frac{[HCO_3^- - x]}{[CO_2 + x]} = pH - 6.38 \quad (2.43)$$

Similarly adding one mole of hydroxide converts one mole of CO_2 into one mole of bicarbonate, such that dosing with y moles of hydroxide gives:

$$\log_{10}\frac{[HCO_3^- + y]}{[CO_2 - y]} = pH - 6.38 \quad (2.44)$$

pH AND ALKALINITY
In a water whose major cationic impurity is Ca^{2+} the relationship between the concentrations of CO_2, HCO_3^-. HCO_3^- and pH are related by the equation:

$$\log_{10}\frac{[HCO_3^-]}{[CO_2]} = pH - 6.38$$

DOSING OF ALKALINITY-LADEN WATERS
Dosing with 1 mole/l of acid generates 1 mole/l of CO_2 and removes 1 mole/l of HCO_3^-.
Dosing with 1 mole/l of hydroxide generates 1 mole/l of HCO_3^- and removes 1 mole/l of CO_2.
This changes the pH in accordance with Equation 2.43.

2.5.3 Buffering

BUFFERING
Both HCO_3^- and CO_3^{2-} will "mop up" any added acid and produce CO_2 with only a modest drop in pH until both are completely used up.

Buffering describes the effect produced by certain chemical systems in water when acid or alkali is added. These systems can suppress changes in pH by reacting to reduce the change in acid concentration. The carbonate system is an example of a buffer, and its action can be best described with reference to equation 2.43 and to a specific water analysis, such as that given in Table 2.5.

Table 2.5 – Typical groundwater analysis, pH 7.5

Cation	Concentration mg/l	Concentration mmol/l	Anion	Concentration mg/l	Concentration mmol/l
Ca^{2+}	100	2.5	HCO_3^-	270	4.4
Mg^{2+}	5	0.2	Cl^-	50	1.4
Na^+	35	1.5	SO_4^{2-}	45	0.5
			NO_3^-	10	0.2

At pH 7.5 Equation 2.41 predicts that the water contains about 0.3 mmol/l of dissolved CO_2. If HCl is added to this water, the increase in H^+ ions drives the first dissociation stage of Equation 2.35 backwards. The concentration of HCO_3^- falls and is replaced by Cl^- whilst the concentration of dissolved CO_2 rises. Almost all the H^+ ions are consumed in the reaction: of those added in the form of HCl, only a few remain.

Adding one mmol/l of HCl to pure water would reduce the pH to around 2, but adding 1 mmol/l of HCl to the groundwater reduces the HCO_3^- to $(4.4 - 1) = 3.4$ mmol/l and increases the dissolved CO_2 to $(0.3 + 1) = 1.3$ mmol/l. Using Equation 2.43 the new pH can be calculated as 6.8, a far smaller drop in pH than one would expect from the quantity of acid added.

Acidifying any water which contains HCO_3^- only leads to a small change in pH, which can be predicted using Equation 2.43. This effect is called *buffering.* When the amount of acid added is equivalent to the HCO_3^- in the water, the pH will be 3.8 and no HCO_3^- remains: the water is now unbuffered and further acid doses will yield the same pH as in pure water. The same kind of mechanism applies if alkali is added to water containing dissolved CO_2, which it converts to HCO_3^- without causing a significant rise in pH.

2.5.4 Hardness

HARDNESS
Calcium bicarbonate decomposes on heating to form CO_2 and $CaCO_3$, which precipitates. Calcium bicarbonate is often described as *Temporary Hardness or Carbonate Hardness.*

In many applications, the most important salts in water are the sparingly soluble salts of calcium and magnesium which tend to precipitate as tenacious crystalline deposits in boilers, cooling towers and heat exchangers and form scum in commercial laundries and in food processing and soft drinks manufacturing. Because of their ability to form scale and scum, calcium and magnesium salts are called *hardness salts.*

Calcium is a major constituent of the earth's minerals – especially as $CaCO_3$, which is the main constituent of chalk or limestone. Rainwater is slightly acid due to its dissolved CO_2 (even without man-made acid rain). On passing through minerals in the ground it will re-dissolve $CaCO_3$ to form Ca^{2+} and HCO_3^- ions. Rainwater which percolates through the ground starts out acidic but usually ends up at pH between 7.5 and 8, with Ca^{2+} and HCO_3^- as its main impurities.

The properties of Ca^{2+} and Mg^{2+} are very similar, and in most natural water the concentration of Mg^{2+} is much smaller than that of Ca^{2+}. Water analyses often lump them together as "Hardness, mg/l as $CaCO_3$". Hardness is often troublesome in industry because it forms insoluble scale in boilers and cooling systems.

If the water having the analysis shown in Table 2.5 is boiled, the partial pressure of all gases in the vapour phase is reduced to almost zero. All gases dissolved in the water, including the CO_2, then come out of solution and are swept away by steam. If no dissolved CO_2 is available, the first dissociation stage of Equation 2.35 can only go backwards, with HCO_3^- ions taking up H^+ ions and reverting to H_2O and CO_2. The CO_2 which is produced in this way is continuously taken away also, so that no equilibrium can be established. H^+ ions are therefore consumed and the pH rises. When it reaches 8.3, the second stage dissociation becomes significant and the CO_3^{2-} which forms $CaCO_3$ with any Ca^{2+} in the water and precipitates, which is how a scale or *"fur"* forms in kettles and pipes. The overall reaction can be written:

$$\mathbf{Ca(HCO_3)_2 = CaCO_3 + CO_2 + H_2O} \qquad \mathbf{(2.45)}$$

When water is heated, the HCO_3^- will cause the equivalent amount of Ca^{2+} and/or Mg^{2+} to precipitate. The amount of hardness which will precipitate in this way is called the *Temporary Hardness* or *Carbonate Hardness*. It is the amount of hardness or bicarbonate in the analysis, whichever is the lower. In most waters (such as the typical analysis of Table 2.5) the HCO_3^- content in equivalents per litre is lower than the hardness. The hardness which is left after that is called the *Permanent Hardness or Non-carbonate Hardness*.

When lime is added, hydroxide ions combine with protons and disappear as water. As the pH rises, and the water contains more and more Ca^{2+} and HCO_3^-. These remain dissolved because calcium bicarbonate $Ca(HCO_3)_2$ is quite soluble. $CaCO_3$, on the other hand, is very insoluble, and precipitates whenever Ca^{2+} and CO_3^{2-} ions are present together at more than trace levels. If natural water containing the usual amounts of Ca^{2+} is brought to pH 8.3 or above, it will precipitate $CaCO_3$. This is the basis of *Clarke's Process* (or lime softening), one of the earliest water treatment processes. Lime softening works in accordance with the following equation:

$$\mathbf{Ca(HCO_3)_2 + Ca(OH)_2 = 2CaCO_3 + 2H_2O} \qquad \mathbf{(2.46)}$$

and, by removing both hardness and alkalinity it also reduces the total dissolved solids content of the water. Note that lime softening alone will only remove only the temporary hardness. Permanent hardness can also be partially removed by the addition of sodium carbonate (soda ash), which effectively converts non carbonate hardness into carbonate hardness and sodium salts, for example with $CaSO_4$:

$$\mathbf{CaSO_4 + Na_2CO_3 = CaCO_3 + Na_2SO_4} \qquad \mathbf{(2.47)}$$

Because lime softening precipitates hardness as calcium carbonate it was important for designers to know how much sludge would be produced and Clarke expressed concentrations *as $CaCO_3$*, a convention which is still often used in water analyses.

SCALE

If there is more than a trace of both Ca^{2+} and CO_3^{2-} in a water they will precipitate $CaCO_3$. This will take place in almost all natural waters if the pH is raised above 8.3.

Such a water will also precipitate $CaCO_3$ if it is boiled or heated for long enough for significant amounts of CO_2 to be lost from it.

2.5.5 Langelier Saturation Index

LSI
The LSI predicts whether a water will be "scale forming" (LSI +ve) or "corrosive" (LSI −ve).

$$LSI = pH - pH_s$$
$$pH_s = pCa + pHCO_3^- + C$$

For a water that is just saturated with calcium carbonate (that is, it will neither deposit nor dissolve scale) Equations 2.26 and 2.40 can be rearranged to give:

$$\mathbf{pK_{SP} = pCa + pCO_3} \qquad \mathbf{(2.48)}$$

$$\mathbf{pK = pH + pCO_3 - pHCO_3} \qquad \mathbf{(2.49)}$$

Combining these two equations by eliminating pCO_3 gives:

$$\mathbf{pH = pK_{SP} - pK - pCa - pHCO_3} \qquad \mathbf{(2.50)}$$

This calculated value of pH is called the *saturation pH* and is designated as pH_S. The difference between the actual pH of the water and its theoretical pH_S value ($pH - pH_S$) is called the Langelier Saturation Index and indicates whether the water will deposit calcium carbonate (*"scaling"*) or dissolve calcium carbonate (*"corrosive"*). A positive LSI value indicates a scaling water and a negative LSI a corrosive water.

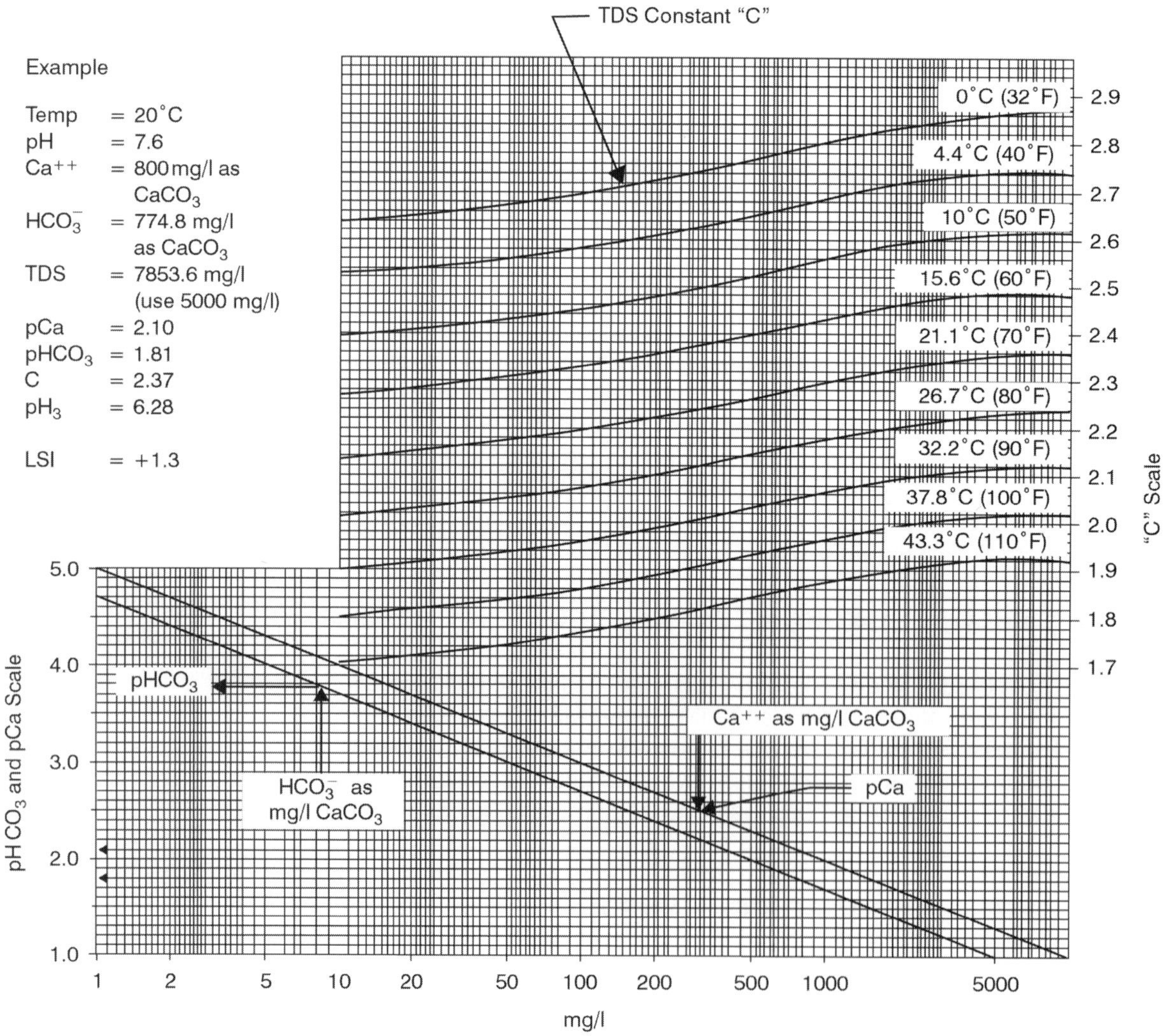

Fig. 2.2 – Langelier saturation index nomogram

The values of pK_{SP} and pK from Tables 2.1 and 2.3 respectively are 8.06 and 6.38. In fact they vary with temperature and total salinity and LSI is usually calculated from a nomogram like that shown in Figure 2.2.

The Langelier Saturation Index is particularly useful for predicting scaling in conditions where a water becomes concentrated as in an atmospheric cooling tower or in a reverse osmosis desalination plant. It becomes unreliable in high salinity waters like, for example, sea water and here it is necessary to use an empirical version such as the Stiff and Davis Saturation Index.

2.6 Self Assessment Questions

SAQ 2.1 If 36% of some micro-organisms die off within 10 hours, what is their half life and how long would it take for 99% of them to die off, assuming first order kinetics?

SAQ 2.2 If a species undergoes first order decay with a half-life of 45 minutes, what percentage is removed over 3 hours?

SAQ 2.3 The ion exchange reaction:

$$\mathbf{Na^+ + RH \Leftrightarrow RNa + H^+}$$

has an equilibrium constant K = 1.5.

If water containing 5 mmol/l of sodium ions is treated with a resin which is 80% in the hydrogen form (that is [RH]/[RNa] = 80/20), what is the concentration of sodium in solution after treatment?

SAQ 2.4 A mains water supply has an alkalinity of 280 mg/l as HCO_3^- and a pH of 8.1. What dose in mg/l of hydrochloric acid is needed to reduce the pH to 7.0?

SAQ 2.5 The water in Table 2.5 is to be softened by lime softening. How many kgs of lime must be added to treat 50 m^3 of the water and how many kgs of calcium carbonate will be precipitated?

SAQ 2.6 The rate at which soluble iron undergoes oxidation by air is given by:

$$\mathbf{r = k\,[O_2]\,[Fe^{2+}]\,[OH^-]^2}$$

The water is to be treated by saturating it with air at 1.5 bar under conditions whereby the ratio of the concentration of oxygen in the water and air does not change (i.e. Henry's Law can be assumed to apply). If lime dosing maintains the pH at 7.8 throughout the reaction, what residence time would be required to reduce the iron content to 100 ppb for a k value of 2.12×10^{13} $mol^3/(s.l^3)$?

SAQ 2.7 Calculate the LSI at 10°C of the water in Table 2.5. What is the new value after dosing with 35.5 mg/l of HCl?

2.7 Nomenclature

H	Henry's Law constant
k	rate constant
K	equilibrium constant
K_a	acid dissociation constant
K_{SP}	solubility product
P	total pressure
p_i	partial pressure of ith species
pK	$-\log_{10}K$
pX	$-\log_{10}[X]$
r	reaction rate
[X]	molar concentration of X
x_i	mole fraction of ith species in the liquid phase
y_i	mole fraction of ith species in the gas phase

2.8 Solutions to Exercises

Ex 2.1 $C/Co = 0.75$,
From rate equation, $k = -(1/t) \ln(C/Co)$,
So at $t = 5$, $k = 0.0575\,min^{-1}$

Ex 2.2 $[NH_4^+]/[NH_3][H^+] = 10^{9.3} = 2 \times 10^9$
At pH $< \log(2 \times 10^9) - 2$, $[NH_3]$ less than 1% of total ammonia concn.
Assuming all ammonia as NH_4^+, initial $[NH_4^+] = 21.4 \times 10^{-3}$
So initial $[NH_3]$ at $[H^+] = 10^{-3}$ is 1.1×10^{-8}, 0.18 ppb as NH_3, **0.15 ppb "as N"**.

Ex 2.3 $K_{SP} = [Ca^{2+}][F^-]^2 = 3.2 \times 10^{-11}$
$[Ca^{2+}] = 120\,mg/l = 120/40\,mmol/l = 0.003\,mol/l$
$\therefore [F^-]^2 = 3.2 \times 10^{-11}/0.003 = 1.07 \times 10^{-8}$
$\therefore [F^-] = 1.032 \times 10^{-4}\,mol/l$ = **1.96 mg/l**

Ex 2.4 If pH = 9.5 then pOH = (14 − 9.5) = 4.5
So $[OH^-] = 10^{-4.5} = 3.16 \times 10^{-5}\,mol/l = 0.0316\,mmol/l$
The mol weight of NaOH = 40 so the concentration is (40 × 0.0316) = **1.26 mg/l**.
1 mole HCl = 1 + 35.5 = 36.5 g
1 mg/l acid equates to $10^{-3}/36.5\,moles/l = 2.74 \times 10^{-5}\,M$
1 mole H^+ reacts with 1 mole OH^- to produce 1 mole water
Thus, new $[OH^-] = 3.16 \times 10^{-5} - 2.74 \times 10^{-5} = 4.2 \times 10^{-6}\,M$
So, new pOH $= -\log(4.2 \times 10^{-6}) = 5.4$
And so new pH = 14 − 5.4 = **8.6**

Ex 2.5 Henry's constant for chloroform $H_{CHCl_3} = 2.55 \times 10^{-5}\,l.atm/mg$
$CHCl_3$ concentration in solution $= 12\,\mu g/l = 12 \times 10^{-3}\,mg/l$
$p_{CHCl_3} = Hx = 2.55 \times 10^{-5} \times 12 \times 10^{-3}$ = **3.06×10^{-7} atm**

Ex 2.6 From Table 2.1 $H_{air} = 0.046\,l.atm/mg$
At 1 atm $x = 1/0.046 = 21.7\,mg/l$
At 4 atm $x = 4/0.046 = 86.9\,mg/l$
$\therefore$ air released = (86.9 − 21.7) = **65.2 mg/l**

Ex 2.7 The free CO_2 concentration is determined by Henry's Law.
From Tables 2.2 or 2.3, $H_{CO_2} = 2.72 \times 10^{-3}\,atm.l/mmol$
At 1 atm the p_{CO_2} is $(1 \times 0.03/100) = 3 \times 10^{-4}$ atm and
$x (=[CO_2]) = 3 \times 10^{-4}/2.7 \times 10^{-3} = 0.11\,mM$, or $1.1 \times 10^{-4}\,M$
Eqn 2.39 gives: $[H^+][HCO_3^-]/[CO_2] = 4.14 \times 10^{-7}$
Eqn 2.35 shows that $[H^+] = [HCO_3^-]$ for CO_2 dissolution in pure water, and so:
$[H^+]^2/1.1 \times 10^{-4} = 4.14 \times 10^{-7}$
$\therefore [H^+]^2 = 1.1 \times 10^{-4} \times 4.14 \times 10^{-7} = 4.55 \times 10^{-11}$
$\therefore [H^+] = 6.75 \times 10^{-6}$ so pH = **5.2**.
Molar mass of NaOH = 40.
So, 2 mg/l NaOH equates to $2/40\,mM = 5 \times 10^{-5}\,M = y$ in Equation 2.44.
Thus, new pH
$= \log[(6.75 \times 10^{-6} + 5 \times 10^{-5})/(1.1 \times 10^{-4} - 5 \times 10^{-5})] + 6.38$
= −0.024 + 6.38 = **6.36**.

Ex 2.8 P = 120, M = 290.
P < ½M so $CO_3^{2-} = 2P = 2 \times 120 = 240\,mg/l\ CaCO_3$.

3. Colloid and Surface Chemistry

Contents

Aims and Objectives

This Unit covers the chemistry of colloids and surfaces hydraulics of particles moving in water. Separation of particles from water is used in almost every water treatment system so it is important to understand the principles.

After studying these notes you should be able to:

1. explain in your own words the following terms and concepts:

   colloid
   hydrophobicity
   diffusion
   electrical double layer
   colloid stability

It is not essential for you to memorise any equations or constants.

It is important that you are able to correctly complete all the self assessment questions at the end of these notes.

Essential Prerequisites

Prior to undertaking this lecture it is important that you have completed and understood the following Unit from the *Process Science and Engineering* Module:

Unit 7 Fundamentals of Process Engineering

3.1 Introduction

Many impurities in water and wastewater are insoluble suspended particles. One particular class of insoluble contaminants, *colloids*, cause particular problems of *haze* and *turbidity*. Colloids are very fine particles which carry an electrostatic surface charge which stabilises them so that it is difficult to determine whether they are actually suspended or dissolved. Indeed some large organic *macromolecules* exhibit many characteristics of colloids.

A common example of a colloidal system is fog and this exhibits a characteristic common to all colloids, namely that of scattering light shone into it. The effect of fog on car headlights is well known. This so called *Tyndall effect* should not to be confused with *absorption* of light which occurs in coloured solutions. If a light is shone through a coloured solution then the total amount of light leaving the solution is less than that entering. In the case of a turbid or hazy water all the light which enters the water leaves it but in a more diffuse and multidirectional manner.

The measurement of turbidity by light scattering techniques is called *nephelometry* and the most widely used units of measurement are *nephelometric turbidity units* (NTU) which compare the turbidity of the water with that of a known standard solution of *formazine*.

Section 3.2 requires a study time of about 1 hour

3.2 Particles

3.2.1 Suspended Solids

Many impurities in water and wastewater are in the form of suspended particles. It will become apparent in Unit E12 that, once quiescent or semi-quiescent conditions occur, these particles will settle under the influence of gravity. The larger and denser the particle the faster it will settle whilst smaller, lighter particles will tend to remain in suspension for a longer period.

Fast flowing rivers and sewage streams will carry silt and grit which are visible to the naked eye as discrete particles but suspended solids of much smaller sizes are also found as the chart in Fig. 3.1 shows.

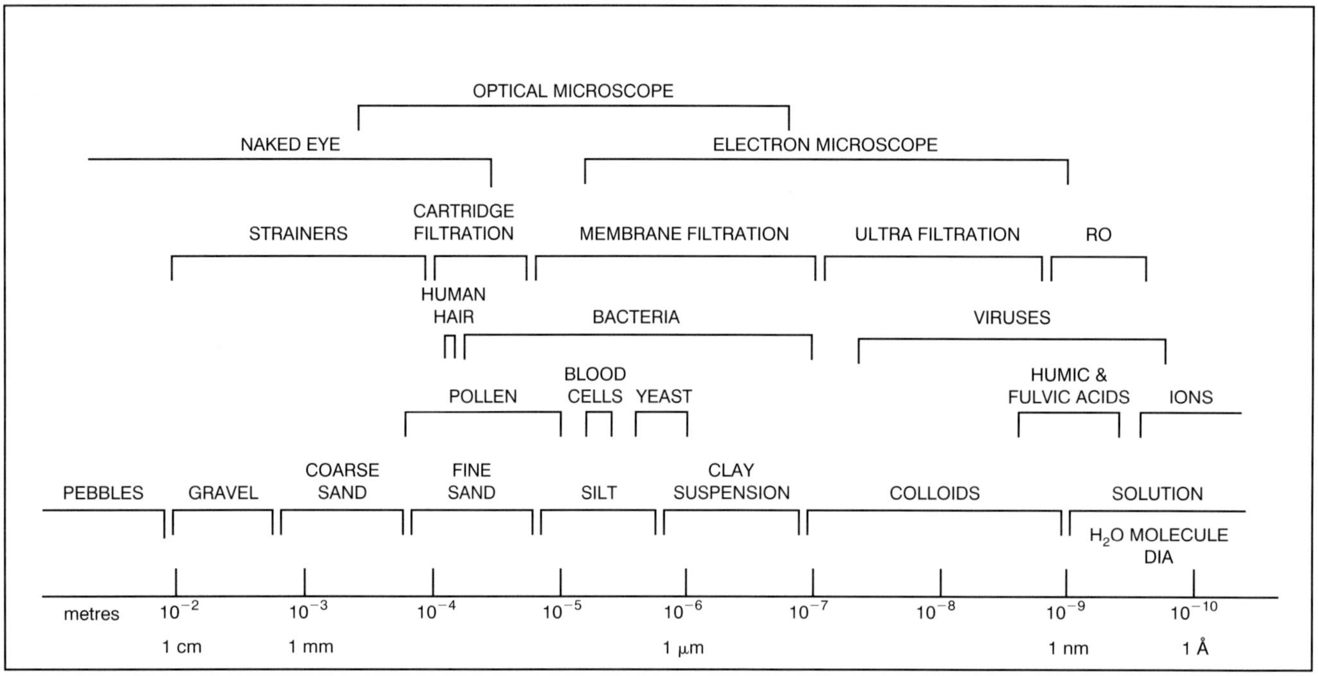

Fig. 3.1 – Particle sizes

COLLOIDS
Colloidal particles are generally considered to be those with diameter less than about 1 μm.

The very smallest particles, having at least one dimension smaller than about 1 micron (1 μm or 10^{-6} m) are called *colloids*. A lower size limit is usually set 1000 times smaller at 1 nm (10^{-9} m), since at this point it is difficult to distinguish between particles and dissolved molecules. In natural waters particle sizes extend up to larger values (hundreds of microns), beyond the colloidal range. The distinctive colloidal properties of particles do not end abruptly when the size exceeds 1 μm – there is a gradual transition over a fairly wide size range.

Some characteristic properties of colloidal particles which, to some extent, justify the distinction between these and larger "suspended" particles are:

a) A relatively large surface area
b) Diffusion coefficients are higher
c) Sedimentation rates are low
d) Difficulty of observation by optical microscopy.

3.2.2 Surface Properties

A solid particle of any size and shape has a distinct mass, volume and surface area. For simplicity, only spherical particles will be considered here, since calculations are very much simpler. For non-spherical and irregular particles it is usual to report their properties in terms of "equivalent" spherical particles. For spheres of diameter d, the simple geometric formulae for volume, V and surface area, A are:

Ex 3.1
If the particle diameter is doubled by what factor is (a) the area, and (b) the volume increased?

$$V = \pi d^3/6 \qquad (3.1)$$
$$A = \pi d^2 \qquad (3.2)$$

It follows that the surface area per unit volume is:

$$A/V = 6/d \qquad (3.3)$$

and the surface area per unit mass of particles of density ρ is:

$$A/m = 6/(\rho d) \qquad (3.4)$$

From this simple result it can be shown that surface area per unit mass (*specific surface area*) can become quite large as the particle size is reduced. With particles of sub-micron diameter values of thousands of square metres per kilogram become possible. Note that for spherical particles, as assumed here, the specific surface is a minimum value – lower than can be achieved by any other shape. A convenient measure of sphericity is based on a comparison of the surface area of a particle with that of an equivalent sphere, having the same mass or volume.

Ex 3.2
Calculate the specific surface area for particles with a density of 2000 kg/m^3 and a diameter of 1 μm. How would the surface area increase if the particle size were reduced to 0.1 μm?

At the surface of a particle in water there is an interface between two phases: solid and liquid. It is possible to consider other interfaces such as liquid–liquid (emulsions) and gas–liquid (bubbles and foams), which are relevant to some aspects of water and wastewater treatment. However most attention here will be paid to the solid–liquid system.

The interface between two phases has a very important property – the *Interfacial Energy* (often called the *Surface Energy*, although this term strictly applies to the interface between one phase and a vacuum). This is the energy associated with creating the interface and is usually quite high. With liquids the surface energy is apparent as *surface tension*, which is the corresponding force. Drops of liquid, such as water, tend to adopt a spherical shape in order to minimize the surface energy. Work has to be done to extend a surface because more molecules have to be brought close to the surface and hence lose some intermolecular contact with their neighbours. Since molecules in one phase have greater affinity for themselves than for molecules in an adjacent, immiscible phase, the creation of an interface is an unfavourable process, which requires energy to be expended. (When molecules in different phases have a greater affinity for each other than for their own kind, then the two phases are miscible, or there is solubility in the solid-liquid case, so that two distinct phases cannot exist and the question of an interfacial energy does not arise.)

Although not directly measurable as in the case of liquids, the surface energy of a solid is very real and has important consequences. Dividing a solid object into fine particles (by grinding, for instance), requires chemical bonds to be broken and, again, energy has to be expended, which appears as surface energy of the particles.

3.2.3 Colloids

The large surface area associated with colloidal particles is one reason why the subject of *Colloid and Surface Chemistry* is often treated as a distinct discipline.

In a liquid such as water finely-dispersed insoluble particles have an interfacial energy and a dispersion of such particles in water is *thermodynamically unstable* in the sense that a lower (free) energy state could be reached if the particles became bound together in aggregates. In this way some particle-particle contacts would be formed at the expense of particle-water contacts, which are less favourable. The fact that particles may not spontaneously coagulate is a result of *kinetic stability*, caused by inter-particle repulsion (see below). Dispersions of this type in water are known as *hydrophobic colloids*. For other liquids the more general term *lyophobic colloids* is used. These terms imply "water-hating" and "solvent-hating" properties respectively. The term *hydrophobic* is used in different ways. In the sense just introduced it implies nothing more than that particles are essentially insoluble in water. However, in another sense it refers to *surface* properties and wettability (see below).

A quite distinct class of colloids is worth mentioning here. These are essentially *soluble* materials, which happen to exist as "particles" of very small dimensions i.e. macromolecules such as proteins, starches, etc. In fact the word "colloid", coined by Thomas Graham in the 1860s derives from the Greek word meaning glue. Such materials in water are known as *hydrophillic colloids*. They have some very different properties from hydrophobic colloids, especially by virtue of the fact that they are *thermodynamically stable*. In natural waters many dissolved organic substances, such as humic acids, fall into this category.

Hydrophobic colloids in water, because of their large specific surface and interfacial energy, are important sites for *adsorption* of substances from water. This process occurs when materials dissolved in water have an affinity for an interface and accumulate in the interfacial region. Adsorption can occur at the air-water interface (for instance surfactants) and on the surface of solid particles. In the latter case organic substances may adsorb because they have some hydrophobic character and adsorption reduces the contact between water and the hydrophobic parts of the molecule. Adsorption may also occur as a result of hydrogen bond formation, van der Waals attraction or electrostatic effects. It has very important consequences for natural water chemistry, since many organic substances are bound to solid particles. Adsorption is also a very important unit operation in water treatment.

3.3 Transport

Section 3.3 requires a study time of about 1 hour

3.3.1 Transport Mechanisms

Particles in water can move in essentially three ways:

a) Convection – movement with the fluid
b) Diffusion – as a result of molecular motion
c) Sedimentation – under the influence of a gravitational field.

In practice, convection is the most effective mechanism in many cases, but we shall here consider only diffusion and sedimentation, since they are relevant to a number of separation processes.

3.3.2 Diffusion

Particles in water are subject to continual bombardment by water molecules, since the latter are in a constant state of agitation as a result of their thermal energy. This is the origin of *Brownian motion*, which can be observed microscopically for particles smaller than a few microns. It is seen as a random sequence of "jumps" by a particle, which cause the particle to move from its original position. These jumps are the composite effects of many invisible motions – a particle is estimated to change direction about 10 million times a second. The motion of a particle by Brownian motion is known as *molecular diffusion* and the displacement x depends on the square root of the time, t:

$$\mathbf{x = \sqrt{2Dt}} \qquad \mathbf{(3.5)}$$

where D is the *diffusion coefficient*, which for spherical particles moving in a fluid of viscosity μ, is given by:

$$\mathbf{D = \frac{kT}{3\pi\mu d}} \qquad \mathbf{(3.6)}$$

k being the Boltzmann constant (1.38×10^{-23} J/°K). Equation 3.6 shows that the diffusion coefficient follows from dividing kT, a measure of the thermal energy, by a term related to the viscous drag on a particle (see below). Thus the diffusion will be more significant for smaller particles, since these have the same average thermal energy as larger particles, but their motion is less restrained by drag.

It is worth pointing out that even quite small colloidal particles are not moved far by diffusion alone. For a particle with a diameter of 0.1 μm, in water at normal temperature, Equation 3.6 gives $D = 4.4 \times 10^{-12}$ $m^2.s^{-1}$, and, from Equation 3.5, it would take the particle nearly 1½ days to be displaced by 1 cm.

Ex 3.3
Calculate the time taken for a 0.1 μm particle to move 0.1mm by diffusion alone at ambient temperature assuming a diffusion coefficient of $4.4 \times 10^{-12} m^2.s^{-1}$.

It is instructive to calculate the time required for a particle to move through a distance equal to its own diameter. In this case, movement by sedimentation and diffusion takes the same time (around 1 second) for particles with a diameter of about 1 micron, which is another reason for the choice of 1 μm as the boundary between colloidal and suspended particles. Larger particles will tend to settle out, whereas sub-micron particles are likely to be less influenced by sedimentation.

3.3.3 Wetting and Hydrophobicity

We have seen that small insoluble particles in water are usually called "hydrophobic colloids". The term "hydrophobic" is also used to describe surfaces which are not easily wettable by water. This can be confusing in the case of substances such as silica, which has only a very low solubility in water. Particles of silica in water can be described as a hydrophobic colloidal system, whereas a silica surface is easily wetted by water and so is not "hydrophobic" in the second sense.

Figure 3.2 shows a drop of water on a plane solid surface. It is subject to a number of forces, which may cause the water to spread over the surface or to remain as a separate drop. In the latter case there is a measurable *contact angle*, θ, which is a measure of the *hydrophobicity* of the surface. The contact angle appears in the *Young Equation*:

$$\gamma_S = \gamma_{SL} + \gamma_L \cos\theta \qquad \textbf{(3.7)}$$

where γ_S and γ_L are the surface energies of the solid and liquid respectively and γ_{SL} is the interfacial energy for the solid-liquid interface.

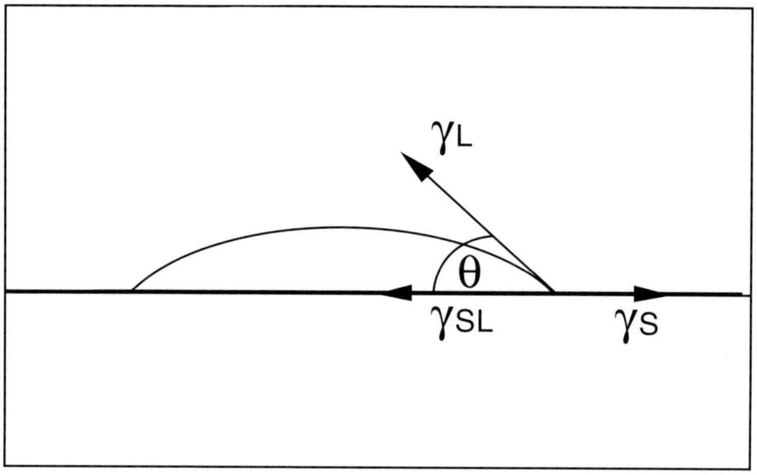

Fig. 3.2 – Contact angle of a drop of water on a solid plane

One way of arriving at this expression is by considering a force balance at the line where the three phases meet. The various γ terms may be regarded as interfacial energies (having units of $J.m^{-2}$) or as interfacial tensions ($N.m^{-1}$). However the Young equation is more convincingly justified in terms of energies. The essential point is that contact angle is determined by the relative magnitudes of the interfacial energies. Rearranging equation 3.7:

$$\cos\theta = \frac{\gamma_S - \gamma_{SL}}{\gamma_L} \qquad \textbf{(3.8)}$$

Ex 3.4
For a contact angle of 45° and an interfacial energy half that of the solid surface energy, how much greater is the solid surface energy than the liquid surface energy?

A low value of θ, directly measured by optical microscopy, can be due to a low value of the liquid surface energy, a high value of the solid surface energy and/or a low value of the solid-liquid interfacial energy. The last condition is achieved when there is significant attraction between the two phases. So, water will have a greater tendency to spread over surfaces which have a high energy and an attraction for water. Good examples are oxides, such as silica, SiO_2, which, in contact with water become hydroxylated to give silanol groups, –SiOH. These form hydrogen bonds with water, leading to a low interfacial energy. Such materials are completely wetted by water and have a contact angle of zero.

By contrast, surfaces such as paraffin wax and PTFE have no bonding sites for water and have fairly low surface energies, so that they are difficult to wet and have high contact angles (around 110°). Teflon forms effective "non-stick" surfaces because of its hydrophobic, non-wetting character. Hydrophobicity plays an important role in the process of flotation, since air bubbles will attach much more readily to hydrophobic surfaces.

Section 3.4 requires a study time of about 1 hour

3.4 Surface Electrical Properties

3.4.1 Surface Chemistry

The surfaces of most particles in contact with water are charged and in natural waters the charge is nearly always negative. There are several reasons for the development of surface charge and a detailed treatment is not possible here. In many cases, some form of *surface ionization* occurs in which dissociation of cations leaves a negatively charged surface. Oxides, for instance, have surface hydroxyl groups in contact with water, as in the case of silanol groups mentioned above. For a metal oxide, such as alumina or ferric oxide, the surface hydroxyls can be represented as –MOH and the acid-base properties of these groups influence the nature of the ionization, which can be represented generally as:

$$\mathbf{MOH_2^+ \Leftrightarrow MOH \Leftrightarrow MO^-} \qquad \textbf{(3.9)}$$

The equilibrium is shifted to the right (more negative surface) by the loss of protons, H^+ and hence by an *increase* in pH. Conversely, increasing the concentration of H^+ or lowering the pH causes the surface to become more positively charged. For a given oxide, there is a certain pH value at which there are equal numbers of positive and negative surface sites and a net surface charge of zero. This is the *Point of Zero Charge (PZC)*, which depends on properties of the oxide, especially its acid–base properties. Some typical, approximate, PZC values are given in Table 3.1. This table shows that silica particles tend to be negatively charged in environmental waters (i.e. pH ~ 7) whereas alumina tends to be positively charged.

Table 3.1 – PZC for common metal hydroxides

Oxide	Formula	PZC (pH)
Silicon dioxide (silica)	SiO_2	2
Titanium dioxide (titania)	TiO_2	6
Ferric oxide	Fe_2O_3	7
Aluminium oxide (alumina)	Al_2O_3	9
Magnesium oxide (magnesia)	MgO	12

For an acidic oxide, such as silica, there is a strong tendency for the surface to lose H^+ and silica surfaces are negatively charged at all pH values down to about 2. Conversely, a highly basic oxide, such as MgO, only becomes negatively charged at high pH values. Oxides of intermediate (amphoteric) properties, such as titania and ferric oxide, have PZC values around neutral pH, so that both positive and negative surfaces may be expected under ordinary conditions.

Other types of particle have surfaces with different ionization equilibria. For instance, for biological surfaces the most important ionizing groups are carboxylic acid (–COOH) and amine ($-NH_2$), which ionize to give negative and positive sites respectively. Again, there is a characteristic pH value at which the surface has a net zero charge, the PZC. For many such surfaces in natural waters the PZC is rather low (around 4–5), so that at ordinary pH values the surface would be negatively charged.

Particles in natural waters are of various types and of widely different origins, thus a range of particle charges, both negative and positive, might be expected. However, it is usually found that the overwhelming majority of particles are negatively charged. This is partly a result of *adsorption* of organic materials (mainly *humic substances*) from solution on particle surfaces. Humic acids and related materials are negatively charged under most conditions, because of ionization of carboxylic acid groups. This means that positively-charged sites on the underlying particle surface would be neutralized by the adsorbed organic layer and the overall charge becomes negative.

3.4.2 The Electrical Double Layer

A charged surface in contact with an electrolyte solution develops a characteristic distribution of charge between surface and solution. Ions in solution tend to be attracted to an oppositely-charged surface, but this tendency is opposed by the randomizing influence of thermal energy. The simplest description of the double layer is that now known as the *Gouy–Chapman model*. This has features in common with the later Debye–Hückel treatment of electrolyte solutions. The surface charge is assumed to be balanced by an equivalent number of *counterions* in solution, which are distributed in a *diffuse layer*. Considering, for simplicity, a flat surface, it can be shown that the electric potential in solution varies with distance from the surface in a simple exponential manner:

$$\Psi = \Psi_0 \exp(-\kappa x) \qquad (3.10)$$

where Ψ is the potential at a distance x from the surface, at which the potential is Ψ_0.

The potential in solution falls at a rate determined by the *Debye–Hückel parameter,* κ. This term determines the range of interaction between particles in water and is defined by:

$$\kappa = \sqrt{\frac{2FI}{\varepsilon RT}} \qquad (3.11)$$

where I = ionic strength
F = Faraday (see Unit C1)
ε = permittivity of the solution

Ex 3.5
All other bulk properties being equal, what effect on the Debye-Hückel parameter would changing a 1 M solution of sodium chloride to a 2 M solution of calcium nitrate have?

The ionic strength is a measure of the total ionic concentration and is defined as:

$$I = \frac{1}{2}\sum c_i z_i^2 \qquad (3.12)$$

The parameter κ has the dimensions of reciprocal length and $1/\kappa$ is the *Debye length*, which is a measure of the thickness of the diffuse layer adjacent to a charged interface. At a distance much greater than this, the potential in solution has fallen to nearly zero and the influence of the charged surface is no longer felt.

The simple Gouy–Chapman picture of the electrical double layer has been modified in a number of ways, the most important of which is the *Stern model*. This allows for a layer of counterions close to the surface (the *Stern layer*), outside which the potential falls exponentially. The potential at this layer (the *Stern potential*) can be considerably less than the true surface potential and can be greatly influenced by *specifically adsorbed counterions* from solution. Certain materials used in water treatment, such as aluminium and iron salts and other cationic additives adsorb on negatively charged particles and reduce the effective surface potential.

Added salts can thus influence the electrical double layer in essentially two ways:

a) By increasing the ionic strength and decreasing the diffuse layer thickness. This effect can be achieved by *any* salts, which are then said to act as *indifferent electrolytes*.
b) By the *specific adsorption* of counterions which reduce the Stern potential. If added in excess the Stern potential may be reversed (e.g. from negative to positive).

These effects are illustrated schematically in Figure 3.3 and are relevant to colloid stability (see below) and to the process of coagulation in water treatment. Figure 3.3 shows the variation of electric potential with distance from a charged surface in the presence of:

(a) an "indifferent electrolyte"
(b) an electrolyte with specifically-adsorbing counterions which reduce the Stern potential
(c) strongly-adsorbing counterions which reverse the sign of the Stern potential.

The surface potential is Ψ_0 and the Stern layer thickness is δ.

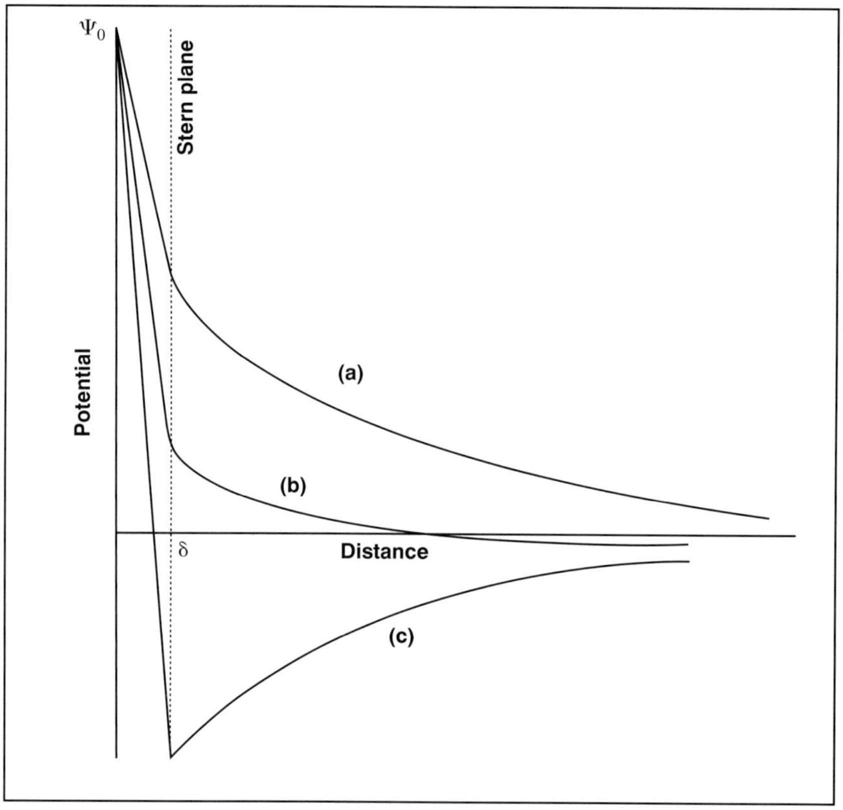

Fig. 3.3 – Electrical potential vs distance from surface, solution ionic strength increasing from (a) to (c)

3.4.3 Electrokinetic Phenomena

Whenever a charged surface and the adjacent liquid are in relative motion, some degree of charge separation occurs. This is because the surface charge and closely-associated ions in the Stern layer are essentially fixed to the surface, whereas the counterion charge in the diffuse layer is free to move with the liquid. The separation of charge gives rise to a number of *electrokinetic* effects, depending on the physical nature of the system and whether the liquid or solid moves. Whatever technique is used, information can be derived on the potential at the *shear plane* between the solid and liquid. This potential is known as the *zeta potential*, after the Greek symbol ζ, used to represent this quantity. The exact location of the shear plane is subject to some uncertainty, but it is generally assumed to lie just outside the Stern layer, so that the zeta potential will be approximately the same or slightly less than the Stern potential.

In some electrokinetic methods, fluid is forced to move relative to the solid and an electric current or potential is measured. One such method is the *streaming potential* technique, which is usually applied to porous media. By forcing liquid to flow though a packed bed of granular media, a potential difference is developed across the bed. This is the streaming potential, which allows the zeta potential of the grains to be calculated. A variant of this method is the *streaming current* technique, in which the flow-generated current is measured rather than the potential. The *Streaming Current Detector (SCD)* is a special device in which the charge carried by particles or macromolecules in water can be investigated. It consists of a cylinder with a reciprocating piston. Charged particles from the water become attached to the walls of the cylinder and piston and the motion generates an alternating streaming current. This method is quite widely used for optimising coagulant dosages in water treatment, which are generally assumed to coincide with the point of charge neutralization and hence to a condition close to zero streaming current.

Another widely used technique is *particle electrophoresis*, in which charged particles move in an applied electric field. By measuring the particle velocity (either by direct microscopic observation or by a laser Doppler method) it is possible to calculate the *electrophoretic mobility* (velocity per unit field strength), from which the zeta potential can be derived. This is the basis of instruments such as the *Zeta Meter*, which can also be used for determining optimum coagulant dosages, although the SCD is more common.

Zeta potentials for particles in natural waters depend on factors such as pH and ionic strength, but generally are found to lie in the range -10 to -30 mV.

Ex 3.6
Which of the parameters cannot be DIRECTLY measured?
(a) contact angle
(b) zeta potential
(c) streaming current
(d) streaming potential
(e) electrophoretic mobility

Section 3.5 requires a study time of about 1 hour

3.5 Colloid Stability

3.5.1 The Nature of Colloids

It has been shown that the so-called "hydrophobic colloids" are inherently unstable, because of their interfacial energy. Aggregation (coagulation) of such particles should occur spontaneously in order to reduce the interfacial area. In very many cases this does not occur because the particles are *colloidally stable*, which means that there is some inter-particle repulsion, hindering contact of the particles. Colloid stability depends on the interplay of various types of interaction between particles (*colloid interactions*). Very brief accounts of the important colloid interactions are given below, without going deeply into theoretical aspects or giving quantitative details.

Before listing some of these interactions, it should be emphasised that they all of comparatively *short range*, that is their range of action is often much less than the particle size. They can play a crucial role when particles are almost in contact but have little influence on particle transport. Also, the magnitude of these interactions is often proportional to the first power of particle size. This is in contrast to other types of interaction such as gravity (proportional to the mass of a particle and hence to the *cube* of particle size) and hydrodynamic drag (proportional to the cross-sectional area of particle and hence to the *square* of particle size). The result is that colloid interactions become progressively more important, relative to other "external" forces, as the particle size decreases. For instance the gravitational force on a 10 micron particle is 1000 times greater than that on a 1 micron particle, whereas colloid interactions would only differ by a factor of 10, for similar conditions.

3.5.2 Van der Waals Forces

The universal attractive forces between atoms and molecules, first recognized by van der Waals, also operate between particles and other objects. Between particles in water there is nearly always a van der Waals attraction, which would allow coagulation to occur, in the absence of any other force. The magnitude of the interaction depends only on the physical properties of the particles and water and, for practical purposes, cannot be significantly changed.

3.5.3 Electrical Repulsion

The fact that particles are usually charged in water, with associated electrical double layers, means that there will be some repulsion between similar particles. This repulsion depends on the surface potential of the particles (usually measured as the *zeta potential*) and the ionic strength of the solution. The latter affects the parameter κ and the Debye length $1/\kappa$ and hence the extent of the diffuse charge layer around the particles. Electrical repulsion between particles becomes significant when their diffuse layers begin to overlap, so that the range of repulsion is influenced by the diffuse layer thickness and hence by the ionic strength. This is illustrated schematically in Figure 3.4. At high ionic strengths (b) the range of repulsion may be reduced sufficiently so that particles can approach close enough for van der Waals attraction to become dominant.

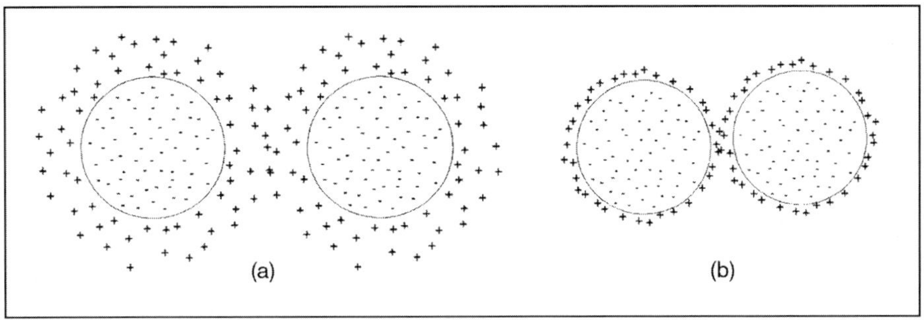

Fig. 3.4 – Interaction of charged particles at a) low and b) high concentration of salt

3.5.4 Steric Interaction

When particles are coated with adsorbed layers of organic material, these layers may act as barriers preventing true contact of the particle surfaces. In the case of polymeric layers, this is usually known as "steric repulsion", because the effect depends on the conformation of adsorbed polymer chains. However in natural waters the effects of adsorbed organics such as humic acids may be largely dependent on their anionic charge and an increased "demand" for cationic additives.

3.5.5 Polymer Bridging

Another effect of adsorbed polymer chains is to bind particles together by a "bridging" mechanism. This effect is exploited in water and wastewater treatment by polymeric flocculants, which will be considered in the Chemical Processes Module.

3.5.6 Combined Interaction – DLVO Theory

The simplest and, so far, the most satisfactory theory of colloid stability considers only van der Waals attraction and electrical repulsion. This theory, developed around 1940 by Deryagin & Landau and, independently, by Verwey & Overbeek, is now usually known as "DLVO theory". Essentially the attraction and repulsion terms are assumed to be additive and are combined to give the total energy of interaction between particles as a function of the separation distance between them. The results of such calculations are shown in Figure 3.5 for 1 μm particles, zeta potentials 20 mV and in a solution of 50 mM 1-1 electrolyte (such as NaCl). The energies due to van der Waals attraction (V_A) and electrical repulsion (V_E) are shown, together with the combined, total interaction energy, V_T. All of the energies are given in kT units (ca. 4×10^{-21} J) for comparison with thermal energy.

When the zeta potential of the particles and the ionic strength are such that repulsion outweighs the attraction, then there is a potential energy barrier, which tends to hinder contact of particles and so prevent aggregation. As the ionic strength is increased and/or the zeta potential reduced (both effects usually result from an increased salt concentration), so the energy barrier becomes lower and contact of the particles can occur more readily. Eventually, the barrier disappears altogether and, in principle, particles can then adhere each time they collide. Such particles are said to be fully destabilized and to undergo rapid coagulation.

Ex 3.7
Which of the following is substantially affected by ionic strength?
(a) ion charge
(b) van der Waal's forces
(c) Stern layer thickness
(d) Debye length
(e) Diffusivity

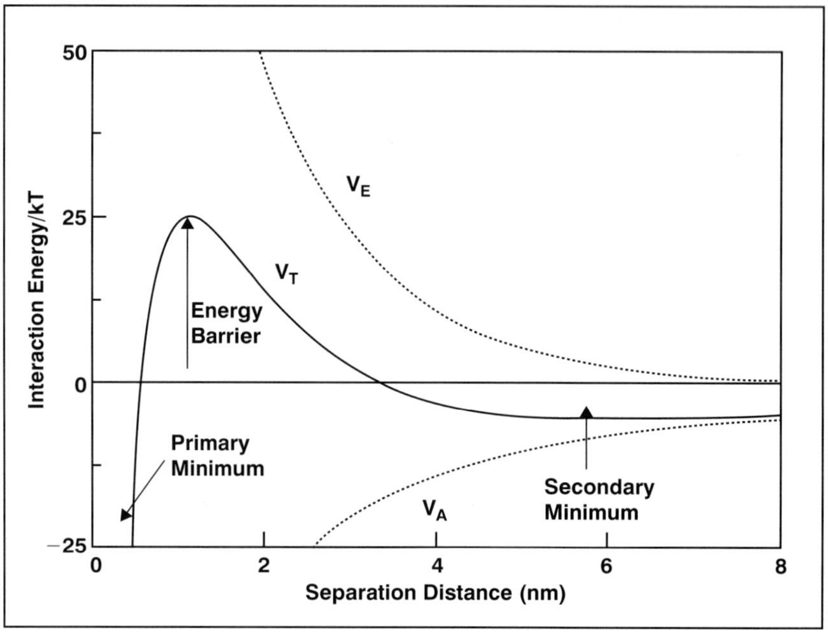

Fig. 3.5 – Interaction energies vs separation distance

There are several features of the potential energy diagram which are important in a discussion of colloid stability. If there is a barrier, then colliding particles must overcome this in order to come into true contact. Potential energies are often expressed in kT units and, in Brownian motion, particles will rarely have an energy of more than a few kT. Thus, a barrier height of 20 kT or more implies a very stable suspension, with only a minute fraction of colliding particles having sufficient energy to surmount the barrier. Once the barrier is overcome, then the particles are held by van der Waals attraction in a deep primary minimum, from which escape would be very unlikely.

The van der Waals and electrical energies depend differently on separation distance and the former is always greater than the latter at sufficiently large separation distances. This gives a *secondary minimum* in the potential energy curve, which can be responsible for the formation of fairly weak aggregates. Since the interaction energies are directly proportional to particle size, secondary minimum effects are more significant with larger particles (greater than about 1 μm in diameter). They are also more important at moderately high salt concentrations, so that the range of electrical repulsion is reduced, allowing the particles to approach closer, where van der Waals attraction is greater.

In the presence of an energy barrier, only a certain fraction of collisions are effective and this fraction is known as the collision efficiency, α. The reciprocal of this value is known as the *stability ratio*, W. By measuring the rate of coagulation as a function of salt concentration, it is possible to find the critical concentration at which the onset of rapid coagulation occurs.

The critical salt concentration, at which rapid coagulation begins, can be reasonably well predicted theoretically and it is known that the counterion charge is of great importance, so that salts with multivalent cations, such as Ca^{2+} and Al^{3+} are more effective at lower concentrations. The strong effect of counterion charge is known as the *Schulze-Hardy rule*.

Although the destabilizing effect of "indifferent" electrolytes (i.e. those without specifically-adsorbing ions) may be reasonably well explained by DLVO theory, in most practical cases we need to consider specific effects of additives. In water treatment practice, coagulation can almost never be achieved simply by increasing ionic strength (although this can have an important influence). It is usually necessary to reduce the zeta potential of the particles in a more direct way. Often, the action of coagulants can only be explained by the adsorption of counterions and consequent reduction of zeta potential. In such cases full destabilization can be achieved at low ionic strength, provided that the zeta potential has been reduced to some critical value. If the additive level is increased, the zeta potential can be reversed and the particles become stable again (a process normally known as *restabilization*).

3.6 Self Assessment Questions

SAQ 3.1 Calculate the surface area of 20 mg mass of particles with a density of 1100 kg/m^3 and a diameter of 0.5 μm.

SAQ 3.2 Calculate the distance moved in 1 hour by a 0.5 μm particle under diffusion through water at a temperature of 285 K

SAQ 3.3 Calculate the ionic strength of a solution containing 5 mmol/l of sodium sulphate and 2 mmol/l of calcium chloride

SAQ 3.4 What would be the effect on the Debye length of increasing the ionic strength of a water from 0.02 to 0.1?

SAQ 3.5 Which of the following salts would you expect to be the most effective and which the least effective in coagulating colloidal silica particles: aluminium sulphate, ferrous sulphate, ferric sulphate?

3.7 Nomenclature

A	area, m^2
d	diameter, m
D	diffusion coefficient, $m^2.s^{-1}$
F	Faraday, 96493 coulombs/eq
I	Ionic strength
k	Boltzmann constant, 1.38×10^{-23} $J.°K^{-1}$
R	universal gas constant, 8.31 $J.°K^{-1}.mol^{-1}$
t	time, s
T	temperature, °K
V	volume, m^3
x	distance, m
z	charge
γ	surface energy, $N.m^{-1}$
ε	solution permittivity
κ	Debye-Hückel parameter
μ	viscosity
ρ	density, $kg.m^{-3}$
θ	contact angle

3.8 Solutions to Exercises

Ex 3.1 (a) quadrupled, and (b) increased by 8 times.

Ex 3.2 Equation 3.4
$\rho = 2000\,kg/m^3$, $d = 1 \times 10^{-6}\,m$
$A/m = 6/(2000 \times 1 \times 10^{-6}) = 3000\,m^2.kg^{-2}$
If $d = 1 \times 10^{-7}$
$A/m = 6/(2000 \times 1 \times 10^{-7}) = 30{,}000\,m^2.kg^{-2}$

Ex 3.3 Equation 3.5
$D = 4.4 \times 10^{-12}$
$x = 0.1\,mm = 10^{-4}m$
$x^2 = 2 \times D \times t$
$10^{-8} = 2 \times 4.4 \times 10^{-12} \times t$
$t = 1136\,s = 1136/60 = 19$ minutes

Ex 3.4 $\cos\theta = 0.707$ at 45°, $\gamma_{SL} = 0.5\gamma_S$
So, $0.5\gamma_S = 0.707\gamma_L$, $\gamma_S = 1.41\gamma_L$ (i.e. 41% greater).

Ex 3.5 I = half sum of cz^2; $\kappa \propto \surd I$: So, $\kappa \propto \surd\Sigma c \times z^2$
$\surd\Sigma cz^2$, 1 M NaCl $= \surd(1^1 + 1^1) = 1.41$
$\surd\Sigma cz^2$, 2 M $Ca(NO_3)_2 = \surd(2 \times 2^2 + 2 \times 2 \times 1^1) = 3.46$
Ratio = 2.46

Ex 3.6 Only the zeta potential (b) cannot be directly measured. It is inferred from other electrokinetic measurements

Ex 3.7 All the parameters listed other than the Debye length are essentially physical parameters unaffected by the total charge in the solution. The ion charge itself is a property only of the intraatomic structure (i.e. the number of electrons vs. the number of protons – see Unit 1).

4. Fundamentals of Microbiology

Contents

Aims and Objectives

This Unit provides an introduction to microbiology which impacts on all aspects of water and wastewater treatment.

After studying these notes you should be able to:

1. explain in your own words the following terms and concepts:

   protists
   pathogens
   doubling time
   specific growth rate
   Monod Equation
   yield coefficient
   hydraulic retention time
   specific oxygen uptake rate

2. carry out calculations for predicting:

   hydraulic retention time
   biomass production

3. recognise

   bacterial and fungal shapes

It is not essential for you to memorise any equations or constants.

It is important that you are able to correctly complete all the self assessment questions at the end of these notes.

Essential Prerequisites

It is not necessary to have completed any other Units before undertaking this Unit.

4.1 Introduction

Microbiology is the study of living organisms that are so small they cannot be seen easily by the naked eye. Most micro-organisms have a diameter of 0.1 mm or less.

Disciplines include algology, mycology, bacteriology, virology and protozoology. Related disciplines include biochemistry, immunology, genetics, ecology and fermentation technology.

This Unit introduces the subject of microbiology and the factors which affect the growth of micro-organisms.

An understanding of micro-organisms is important in most aspects of water and wastewater treatment including:

- public health aspects of water supply and sanitation
- the spread of disease from evaporative cooling towers and air conditioning systems
- biological fouling of pipework and heat transfer surfaces
- water quality for food and drink manufacture
- water quality for pharmaceuticals production and medical applications like renal dialysis
- the use of micro-organisms to treat domestic and industrial wastewaters
- the use of micro-organisms for the treatment and monitoring of water supply

The role of micro-organisms in wastewater treatment for example includes the removal of carbonaceous BOD (Biological Oxygen Demand), the coagulation of non-settleable colloidal solids and the stabilisation of organic matter. The micro-organisms are therefore used to convert dissolved organic material for example into various gases and cell tissue.

Section 4.2 requires a study time of about 1 hour

4.2 The Kingdom of the Protists

4.2.1 Taxonomy

Taxonomy is the system of nomenclature which biologists use to identify living things. Under this system all living things are classified into the following sets and subsets:

Kingdom
Phylum
Class
Order
Family
Genus
Species

PROTISTS
All unicellular living organisms belong to the kingdom *Protista*. Many of these protists are utilised in biological water and wastewater treatment processes.

There are only three kingdoms: *Plantae* (plants), *Animalia* (animals) and *Protista*. It is in this latter kingdom that we find all living organisms that are either unicellular or consist of a group of undifferentiated cells; i.e. all cells look the same. It is a moot point among taxonomists whether viruses should or should not be included (see Section 4.2.3).

Classical taxonomy would classify the protists something like the structure shown in Figure 4.1. However, this is not a particularly useful way of viewing these organisms if we are to make use of them in engineering processes. The classification according to carbon and energy requirements outlined in Section 4.3.1 will be much more useful.

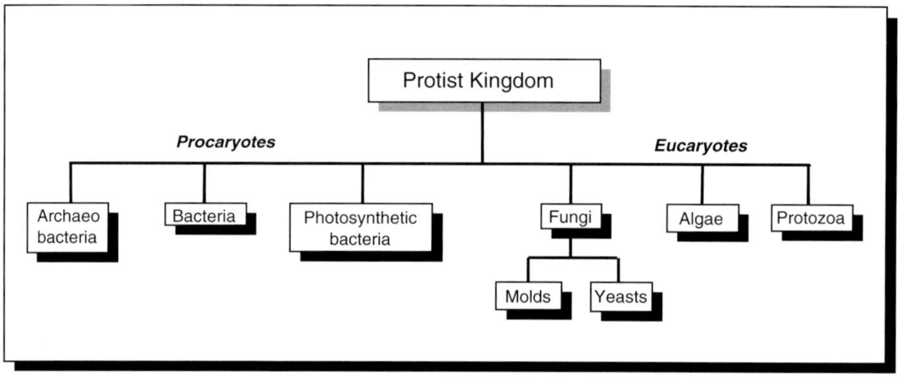

Fig. 4.1 – Classical taxonomy of protists

There are many different species of micro-organisms in the protist kingdom just as there are of plants and animals and it is important to be precise when referring to them so as to avoid any confusion. There are, for example, over thirteen thousand known species of bacteria and over a hundred thousand fungi.

The rules governing the form of nomenclature of micro-organisms can be summarised as follows:

1. The **genus name** comes **first** beginning with a **capital letter**, followed by the **species name** beginning with a **lower case letter**. Examples are: *Vibrio natriegens* and *Thiobacillus ferrooxidans*
2. The name should always be underlined when **handwritten** to indicate italics if printed so "Nitrosomonas europae" in manuscript becomes "*Nitrosomonas europae*" in print

3. After the **full species name** has been used once in a text it is acceptable, thereafter, to use the **abbreviated form,** for example *A. niger* for *Aspergillus niger* and *B. subtilis* for *Bacillus subtilis*
4. If referring to **more than one species of the same genus** this can be indicated by the postscript "*sp*" (eg *Acinetobacter sp*) or by the genus name alone: *Pseudomonas sp* or simply *Pseudomonas*

Ex 4.1
Which of the following is correct?
a) Salmonella Paratyphi
b) salmonella Paratyhphi
c) S. Paratyphi
d) Salmonella paratyphi
e) Salmonella <u>*paratyphi*</u>

4.2.2 Cell Types

There are two main cell types, the simpler being *procaryotes* and the more complex *eucaryotes*, which are the type that mammals and humans have. The main distinction between the two types is that eucaryotic cells contain *organelles*, which are responsible for the internal organisation of the cell, whilst procaryotes, being much simpler, do not. The principal differences in cell structure are illustrated in Figure 4.2 and listed in Table 4.1.

CELL TYPES
Simple cells are called *procaryotes*. More complex cells which contain *organelles* are called *eucaryotes.*

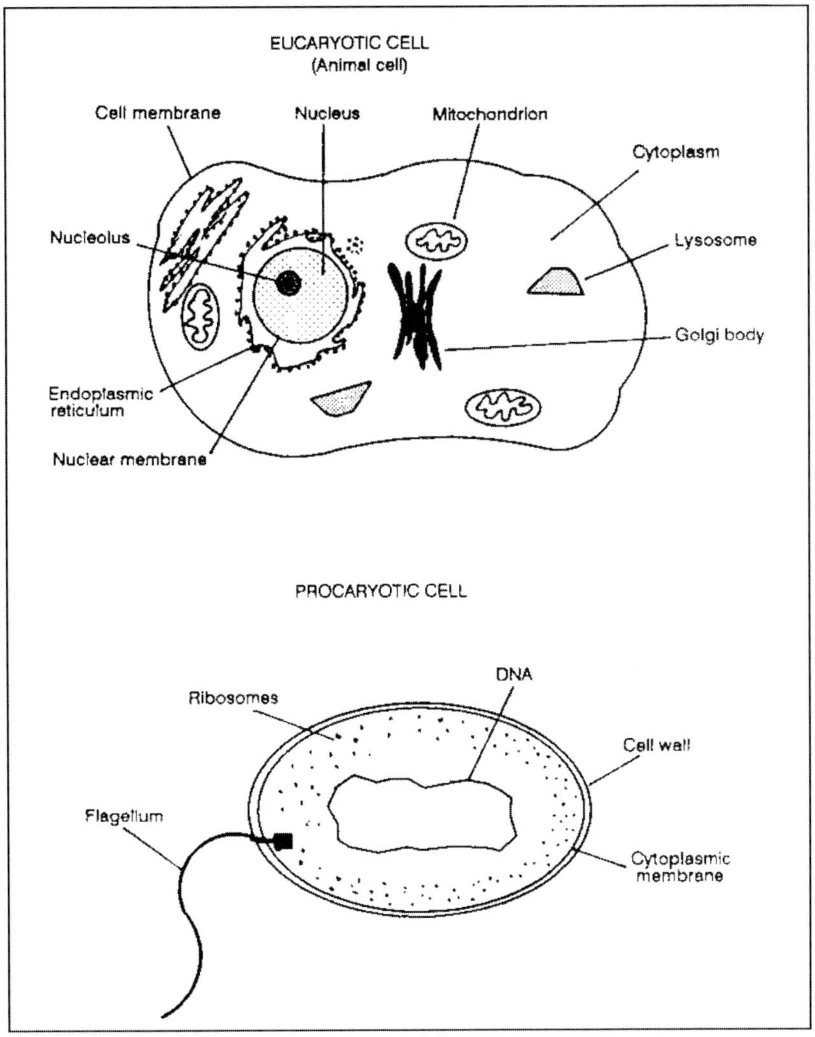

Fig. 4.2 – Microscopic structures within cells

Ex 4.2
Which of the following is not a cell?
a) Organelle
b) Procaryote
c) Autotroph
d) Bacillus
e) Aspergillus

Table 4.1 – Comparison of eucaryote and procaryote

Characteristic	Procaryotes	Eucaryotes
Nucleus	−	+
Number of chromosomes	1	>1
Endoplasmic reticulum	−	+
Golgi apparatus	−	+
Lysosomes	−	+
Mitochondria	−	+
Chloroplasts	−	±
Ribosomes	70s	80s and 70s
Cellulose cell wall	−	±
Peptidoglycan cell wall	+	−

The *archaeobacteria* are very ancient microorganisms, and are often classified as being separate from other procaryotes. In particular, their cell walls have a very different composition. They are of particular interest because many are *extremeophiles*, that is they can grow in very harsh conditions, and include methane producers – methanogens.

Of the other protists, bacteria and protozoa are of most interest as far as water and wastewater treatment processes are concerned. Protozoa are single-celled eucaryotic micro-organisms without cell walls. They are larger than bacteria, in fact they feed on bacteria and are important within wastewater treatment systems, as they help to maintain a natural balance between the different groups of micro-organisms. Protozoa are often able to exist in two different forms: as a cell or as an oocyst, which is similar to the bacterial spore (Section 4.2.5). The cell can be very fragile and in unfavourable conditions, it changes to produce an oocyst, which is hardy and can withstand harsh conditions for quite long periods of time.

Algae are also important in the context of water system microbiology. They are unicellular green plants and use *photosynthesis* (conversion of carbon dioxide and water to sugars using energy from sunlight) as their primary mode of nutrition. They are generally *non-motile* (not able to move independently) and are typically 5–100 μm in size. Algal activity in reservoirs and lakes can profoundly influence the chemistry of the water in which they grow giving rise to seasonal quality variations.

4.2.3 Pathogenic Organisms

PATHOGENS
Micro-organisms which cause disease are termed *pathogenic.*
Pathogens are of two types: *Enteric* organisms inhabit the gut. *Non-enteric* live on our food and cause spoilage.

Most of the bacteria are either useful or benign and only a very few are *pathogens*, that is disease causing. There are two main groups of potentially undesirable micro-organisms: the *enteric pathogens* and the *non-enteric toxin producing bacteria*. In addition to these there are the micro-organisms which cause food spoilage but are not pathogens.

The intestinal tract is a habitat for many types of bacteria, most of which are harmless. However a number of them produce serious diseases including cholera and typhoid, and these are termed enteric pathogens. Examples of human enteric pathogens are given in Table 4.2.

Table 4.2 – Pathogenic micro-organisms

Organism	Disease
Salmonella typhi	Typhoid fever
Salmonella typhimurium	Paratyphoid
Salmonella choleraesuis	Gastroenteritis
Escheria coli	Gastroenteritis
Salmonella schottmulleri	Food poisoning
Vibrio cholerae	Cholera
Shigella dysenteriae	Bacterial dysentery
Legionella pneumophila	Legionnaires' disease

All these bacteria leave the host in excreted faecal matter and enter the next host via the mouth. Disease is then caused by multiplication of the bacteria within the host and production of toxins. Many of these diseases including typhoid, cholera and dysentery are water-borne and result from poor sanitation and, in this respect, both water and sewage treatment are vital to public health. Spores are produced within some bacterial cells. The cell then degenerates, releasing the spore which is a dormant state of a bacteria. These spores are relatively resistant to dryness, lack of nutrients, toxic chemicals, heat and radiation. Therefore true sterilisation processes need to destroy not only live bacteria but these spores also. Another aspect is the control of water used in the production of food and drinks and in the manufacture of pharmaceutical products, where not only total sterility but the removal of bacterial decomposition products is often required.

Some types of bacteria can cause food poisoning without multiplication within the body. These *non-enteric toxin-producing micro-organisms* grow on food and produce a toxin which diffuses into the food. The organism may even be dead by the time the food is consumed but the toxin still active. Examples of organisms and the conditions they produce are given in Table 4.3.

Ex 4.3
Which of the following are not produced by micro-organisms?
a) oocysts
b) hydrogen
c) methane
d) spores
e) toxins

Table 4.3 – Non enteric micro-organisms and disease

Organism	Common source	Effects
Clostridium botulinum	Sausages, canned fish	Botulism
Staphylococcus sp	Chicken	Food poisoning
Bacillus sp	Cooked rice	'Chinese Restaurant Syndrome'
Aspergillus sp	Mouldy peanuts, grain etc	Aflatoxin poisoning
Claviceps purpurea	Rye	'St Anthony's Fire'

Some of the protozoa – *Cryptosporidium* and *Giardia* for example – can cause sickness and so need to be excluded from drinking water. The oocysts are often very resistant even to strongly oxidising biocides like chlorine and others will even withstand boiling for a short time.

Another important group of organisms, as far as public health is concerned, are the viruses. They are neither procaryotes nor eucaryotes, so they are not protists. They are actually obligate intracellular *parasites*, which means that, in order to survive, they require a living host cell that may be procaryotic or eucaryotic. Viruses are usually classified according to the hosts they infect, important classes being animal viruses, algal viruses and bacterial phages. They are also very small (25–350 nm – much smaller than most bacteria) and can only be observed with an electron microscope. They are made up of a core of genetic material, either deoxyribonucleic acid (DNA) or ribonucleic acid (RNA), surrounded by a protein coat which makes them difficult to detect.

Water-borne viruses fall into two main categories: *adenoviruses* which attack their hosts via the bronchial system, and *enteroviruses* which are ingested. Pharyngoconjunctival fever was one of the first adrenoviruses to be identified in the 1950s and was found to be common in swimming baths. Enteroviruses such as the Norwalk virus are responsible for intestinal disorders and are common in rivers in many parts of the world, as are more dangerous viruses like that which causes poliomyelitis.

Virus identification and removal are important aspects of water and wastewater treatment, particularly how they are removed or destroyed by sewage treatment.

4.2.4 Bacterial Morphology

The small physical size of bacteria is important in that it explains some of their properties. In particular, their ability to grow rapidly and the fact that they are at the same temperature as their surroundings is explained by considering the surface area/volume ratio which we discussed in Unit 3. The smaller the volume, the greater the surface area available for exchange of nutrients and heat.

BACTERIA SHAPES
Bacteria occur in three basic shapes:

Spiral – *spirilla*
Spherical – *cocci*
Rods – *bacilli*

Although they are small, micro-organisms can multiply rapidly and have population doubling times of less than 30 minutes under the right conditions. This is considerably shorter than that for any other organism, as Table 4.4 shows.

Table 4.4 – Organism doubling times

Organism	Doubling time
Man	years
Poultry	months
Plants	months
Fungi	hours
Bacteria	minutes

ENUMERATION
Bacteria and fungi are cultured and the resulting colonies are counted to give a number of *colony forming units* per unit volume of water.

Individual species may be identified by staining.

The total number of all bacteria is called a *total viable count* (TVC).

Bacteria occur in three basic morphological types or shapes: spirals (*spirilla*), spheres (*cocci*) and rods (*bacilli*) (Figure 4.3). Note that these Latin descriptive terms are also applied to different species of bacteria e.g., *Bacillus*

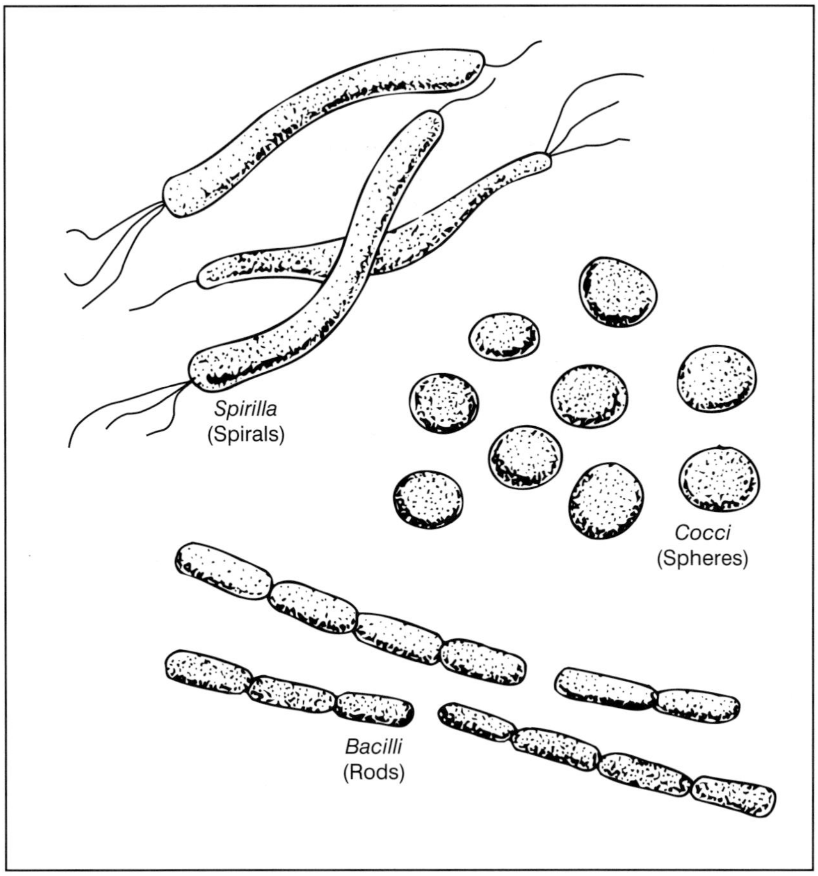

Fig. 4.3 – Bacterial shapes

stearothermophilis, and should not be confused e.g. *Klebsiella aerogenes* is bacillus shaped.

Bacteria and fungi are very often cultured (grown) on agar plates as a first stage of identification and enumeration of bacteria present in a water sample. Bacteria of the same species will grow in clumps known as "colonies" which are visible to the naked eye. These colonies can be distinguished by their different colours, shapes and edge formations as shown in Figure 4.4. Each colony will have started from a single cell so that counting the number of colonies gives an indication of the number of *colony forming units* of bacteria in the original water sample which will normally be given in $cfu\ ml^{-1}$.

Viruses are enumerated by a technique similar to that used for bacteria enumeration, virus numbers being given as *plaque forming units* ($pfu\ ml^{-1}$).

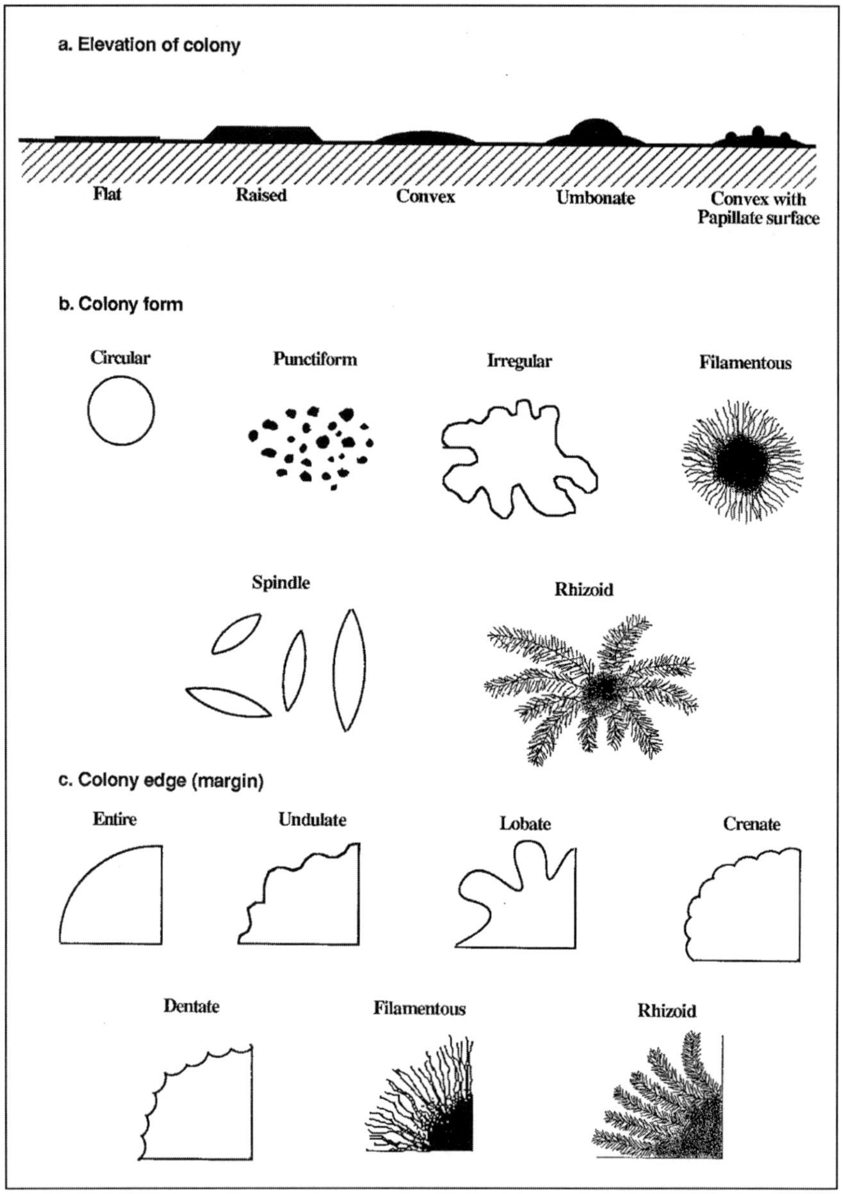

Fig. 4.4 – Descriptions of bacterial colonies

Bacteria of different types take up certain coloured chemicals called stains and this allows identification by colour. The most widely used stain is the *Gram stain*, which differentiates bacteria on the basis of cell wall constituents. There are two types of gram stain: Gram positive (bacteria are stained purple) and Gram negative (bacteria are stained pink). The shape and gram stain of a bacteria form the basis for species identification.

4.2.5 Fungal Morphology

There are essentially 2 main types of fungus as shown in Figure 4.5:

Yeasts are single cells which grow and then divide by binary fission, that is they split into two when they have reached a certain size, while *Filamentous fungi* have long filaments consisting of numbers of undifferentiated cells. They grow by elongation and branching.

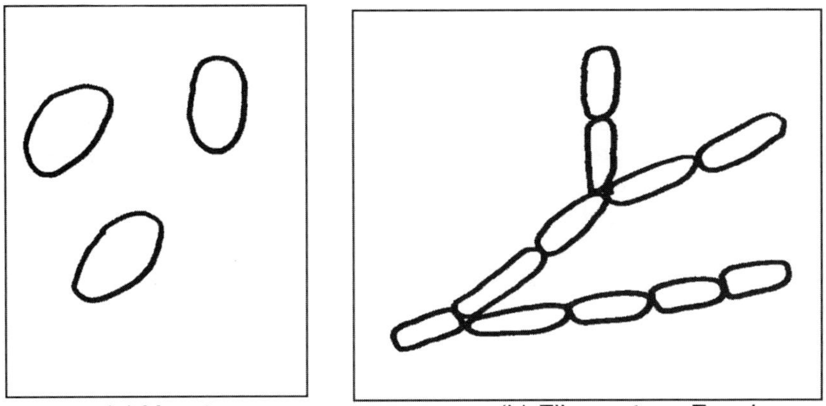

(a) Yeast (b) Filamentous Fungi

Fig. 4.5 – Shapes of fungi

Identification of fungi is mainly based on differences in the reproductive structures.

Section 4.3 requires a study time of about 1 hour.

4.3 Conditions for Microbial Growth

4.3.1 Carbon and Energy Requirements

Probably the most useful way of classifying micro-organisms is on the basis of the way they obtain energy and carbon from which they synthesise cell material. There are two principal groups. *Autotrophic* organisms obtain energy and carbon from inorganic sources whilst *heterotrophic* organisms require organic material as a food source. The suffix *-troph* denotes "feeding habits", and is derived from the Greek word *trophe*, meaning nourishment. As a general rule, autotrophs are less efficient at energy gathering than heterotrophs and therefore grow more slowly.

METABOLISM
Some micro-organisms obtain energy from inorganic sources (*autotrophic*) but some require an organic food source (heterotrophic).

Autotrophs may be *chemotrophic* or *phototrophic.*

Heterotrophs may be *saprophytic* or *parasitic.*

Autotrophic organisms are subdivided into *chemotrophs* and *phototrophs.* Chemotrophs are those organisms that utilise chemical reactions as an energy source – for example sulphate reducing bacteria like *Desulphovibrio desulphuricans* obtain energy by reducing sulphates to sulphides – whilst phototrophs like algae use chlorophyll or similar chemicals to absorb energy from sunlight. Most of the micro-organisms which are used in water and wastewater treatment are chemotrophic.

Heterotrophs are subdivided into *saprophytes* and *parasites*. Saprophytes use dead or decaying plant or animal tissue and are used in the digestion of sewage sludges. Parasites require a living host and many are pathogenic.

4.3.2 Nutrient Requirements

All micro-organisms need O, C, N, H, P and S, which are the elements of organic matter. They also need the monatomic ions of Na, K, Mg, Ca and Cl. Trace elements that might be required include Mn, Fe, Co, Cu, Zn, B, Al, V, Mo, I, Si, Sn, Ni, Cr, F and Se. Some micro-organisms require special growth factors such as vitamins, amino acids etc.

4.3.3 Temperature

TEMPERATURE
Micro-organisms are classified according to their optimum growth temperature.

Psychrophilic 15°C
Mesophilic 35°C
Thermophilic 55°C

Micro-organisms are more robust than most people imagine; they have been found in hot sulphur springs at extreme temperatures and pressures. Indeed, the bacterium *Methanopyrus*, found in submarine volcanoes, has a maximum growth temperature of 110°C and an optimum of 100°C. Different micro-organisms have different temperature profiles and optimum temperatures for growth. There are three groups defined by their temperature range as follows:

psychrophilic organisms	2–30°C
mesophilic organisms	25–45°C
thermophilic	45–70°C

Temperature profiles are shown in Figure 4.6.

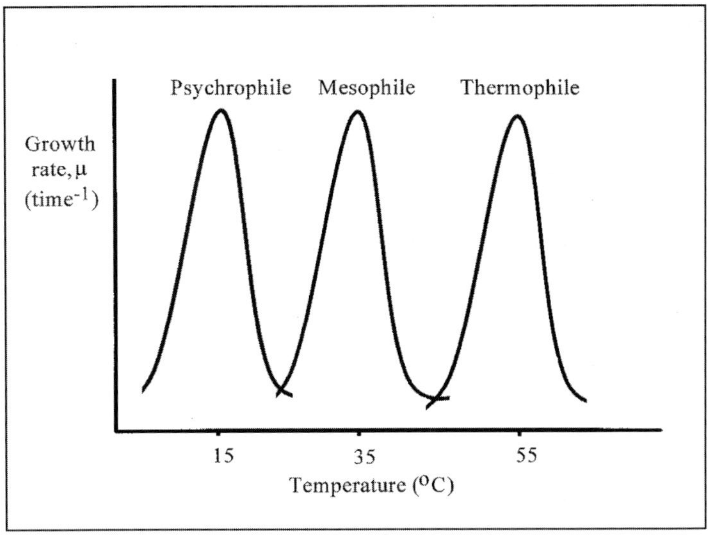

Fig. 4.6 – Temperature profiles for bacterial growth

Ex 4.4
What groups of micro-organisms can use the following substrates as a carbon or energy source?
glucose
carbon dioxide
chlorophyll

4.3.4 pH

Many bacteria grow best at around pH 7, i.e. neutrality. However many species are able to grow at extreme pH values. *Acidophilic* bacteria are adapted to low pH environments, for example *Thiobacillus* can grow at pH 1.0 and their optimum is about pH 1.5–2.0. Very few bacteria can survive alkaline conditions so caustic soda is widely used for cleaning in the food industry, where it provides disinfection as well as fat solubilisation. Figure 4.7 shows typical pH profiles for bacterial growth.

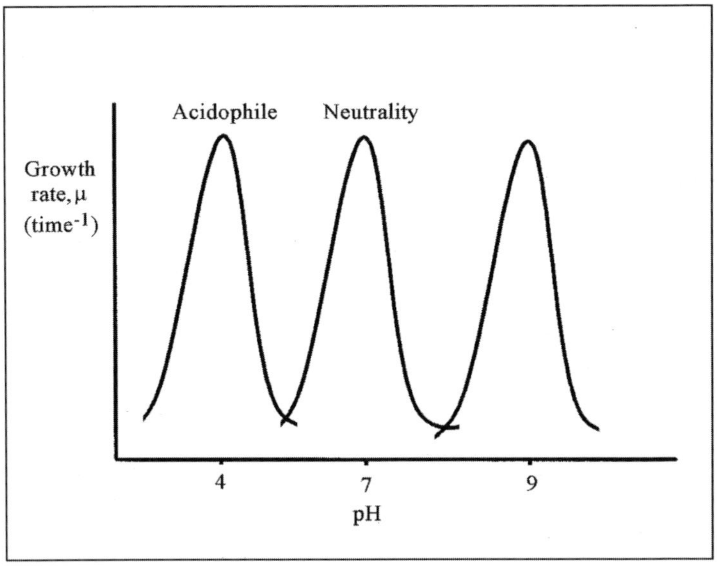

Fig. 4.7 – Optimum pH values for bacterial growth

4.3.5 Oxygen Requirements

OXYGEN
Micro-organisms are classified in accordance with their requirements for oxygen as
Aerobic (need oxygen)
Anaerobic (need absence of oxygen)
Facultative (adapt to conditions).

Micro-organisms differ in their requirements for oxygen. Some are *obligate aerobes*, that is they require the presence of oxygen (or air) in order to live. By far the majority of bacteria, like *Pseudomonas* and *Thiobacillus thio-oxidans* and most fungi are aerobic. At the other extreme are the *obligate anaerobes* which cannot live in the presence of oxygen. Some denitrifying bacteria and sulphate reducing bacteria fall into this category, as do many of the pathogens.

Many bacteria and yeasts are *facultative*, which means that they prefer either aerobic or anaerobic conditions but can adapt their metabolism to the oxygen concentration in their environment and survive in both.

4.3.6 Water

WATER
All micro-organisms require water, although some need more than others.
Most are unable to tolerate high levels of salinity.

All micro-organisms need to be in an environment where water is present in order to grow. Some fungi can grow in relatively dry conditions on solid media but the atmosphere must be damp. Bacteria need to be in an essentially aqueous environment.

Solutes (sugar, salts etc.) dissolved in water have an affinity for water and this is therefore unavailable to micro-organisms. Therefore the more dissolved salts, the less likely it is that bacteria will survive. The availability of water can be expressed as water potential or water activity.

Water activity (a_w) is the ratio of the vapour pressure of a solution divided by the vapour pressure of pure water at the same temperature. Values of a_w range from 0 to 1; 1.000 is pure water. Seawater is 0.980, in which halophile (salt-loving) bacteria live. At a water activity of 0.750 (a *saturated* solution of sodium chloride) only extreme halophiles can survive, e.g. *Halobacterium*. Most micro-organisms can grow at a range of 0.998 to 0.6. Yeasts can tolerate a lower water activity, typically about 0.6, compared with most bacteria that require a water activity greater than 0.98.

4.4 Self Assessment Questions

SAQ 4.1 Which of the following elements cannot be a nutrient?

a) P
b) Co
c) Mo
d) Sn
e) PHe

SAQ 4.2 Which of the following are eucaryotic cells?

a) bacterium
b) yeast
c) human liver cell

SAQ 4.3 Which of the following are pathogenic micro-organisms

a) *Methanosarcina*
b) *Escheria coli*
c) *Methanothrix*
d) *Giardia*
e) *Clostridium*
f) *Cryptosporidium*

4.5 Nomenclature

t = time, h
t_d = culture doubling time, h
μ = specific growth rate, h^{-1}

4.6 Solutions to Exercises

Ex 4.1 *Salmonella paratyphi*

Ex 4.2 Organelle

Ex 4.3 Hydrogen

Ex 4.4 Glucose – Heterotrophic (Saprophytes)
Carbon Dioxide – Autotrophic (Chemotrophic)
Chlorophyll – Autotrophic (Phototrophs)

5. Fundamentals of Biochemistry

Contents

Aims and Objectives

This Unit covers the underpinning biochemical reactions and structures relating to biological processes.

After studying these notes you should be able to:

1. explain in your own words the following terms and concepts:

   biochemicals
   enzymes
   metabolism
   anabolism
   catabolism
   aerobic
   anaerobic
   fermentation

It is not essential for you to memorise any equations or constants.

It is important that you are able to correctly complete all the self assessment questions at the end of these notes.

Essential Prerequisites

Prior to undertaking this unit it is important that you have completed and understood the following units from the *Process Science and Engineering* Module:

Unit 1 Fundamentals of Water Chemistry
Unit 4 Fundamentals of Microbiology

5.1 Introduction

Biochemistry is just that – the chemistry of life. If we have some understanding of the chemical basis of life, it will help in our understanding of the way in which microorganisms function and how they can be used in processing water and wastewaters. Biochemistry is a subject that involves learning a large amount of facts but it can be broken down to simple stoichiometry. This is then of use in mass balances and the design and operation of biological processes.

Section 5.2 requires a study time of about ½ hour

5.2 From Microbiology to Biochemistry

5.2.1 Cells and Chemicals

In Unit 4 we considered the function of living organisms at the level of the cell – which is a closed system in thermodynamic/mass balance terms. In cell biology we look at sub-cellular components and in biochemistry we consider the chemicals that make up life. There is no clear division between the subjects but the hierarchies, based on the size and complexity of the molecules and structures, which are used in the two disciplines are shown in Figure 5.1.

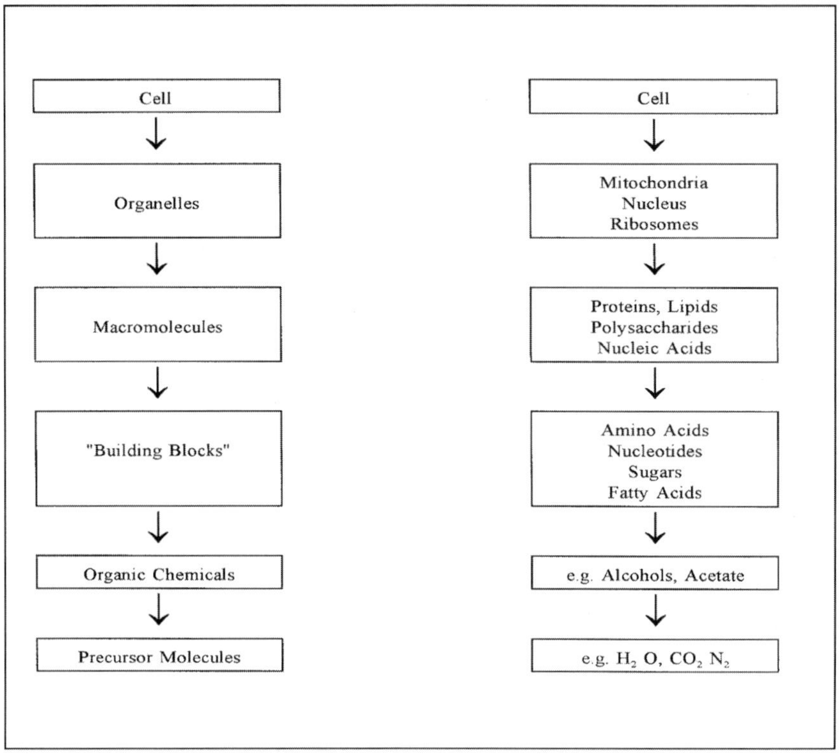

Fig. 5.1 – Relationship between the cell and chemicals

Therefore from microbiology we are studying cells that are too small to see with the naked eye down to biochemistry where we are investigating compounds with molecular weights below 100-or-so. Table 5.1 compares molecular dimensions and weights for some of these biomolecules and cell components:

Components 1 – 3	are the basic building blocks.
Components 4 – 7	are "macro-molecules".
Components 9 – 10	are viruses.
Components 11 and 13	are organelles.
Components 12 and 14	are cells.

Table 5.1 – Biomolecules and cell components

Cell Component	Length (nm)	Molecular Weight
1. Alanine	0.5	89
2. Glucose	0.7	180
3. Phospholipid	3.5	75
4. Myoglobin	3.6	16900
5. Haemoglobin	6.8	65000
6. Myosin	160.0	470000
7. Glutamate dehydrogenase	13.0	1000000
8. Ribosome (*E coli*)	18.0	2800000
9. Phage X174 (of *E coli*)	25.0	6200000
10. TMV (Rod)	300.0	40000000
11. Mitochondrion (liver cell)	1500.0	1×10^{-12} g
12. *E coli*	2000.0	2×10^{-12} g
13. Chloroplast (Spinach leaf)	8000.0	1.3×10^{-10} g
14. Liver cell	20000.0	2×10^{-9} g

Rather than considering a cell to be "merely" a living thing, we can break it down into its component atoms and think of the life processes as a series of chemical reactions. We can use the knowledge of stoichiometry which we obtained in Unit 1 to provide information and formulae for these reactions and, when we understand them, we can use them for process design. Before coming back to this subject, we shall consider some biochemical fundamentals in more detail.

5.2.2 Biochemicals

BIOCHEMICALS
The four main categories of biochemicals are:
carbohydrates, proteins, lipids, nucleic acids.

Biochemical macromolecules are usually put into four main categories:

- *carbohydrates*
- *proteins*
- *lipids*
- *nucleic acids*

There are also complex biochemicals that are mixtures of these or "building block" sized molecules, such as vitamins.

Carbohydrates form structural components and also act as an energy store in cells. Proteins form parts of the cell walls but mainly form *enzymes*, biological catalysts. Lipids are also energy storage compounds but their main function is in forming the semi-permeable cell membrane. Nucleic acids form transport molecules and the genetic material.

Therefore the contents of a cell can be defined in terms of the constituent chemicals as shown in Table 5.2.

Ex 5.1
Comment on the functional importance of the following for a E coli cell:
a) Lipids;
b) Proteins;
c) Carbohydrates;
d) Nucleic acids.

Table 5.2 – Composition of growing cell

Component	% of Total Cell Mass	Av. Molecular Weight	Approx. No. per cell	No. of different kinds
H_2O	71	18	4×10^{10}	1
Inorganic ions (Na^+, K^+, etc.)	1	40	2.5×10^8	20
Carbohydrates	3	150	2×10^8	200
Nucleotides	0.4	300	1.2×10^7	200
Lipids	2	750	2.5×10^7	50
Amino acids	0.4	120	3×10^7	100
Other small molecules	0.2	150	1.5×10^7	200
Proteins	15	40000	10^6	$\approx$2500
Nucleic acids	7	$25000–2.5 \times 10^9$	4.6×10^5	1040

As can be seen, the major constituent is water! Also, it should be noted that the most numerous constituents are proteins, with approximately 2,500 different types present.

5.2.3 Enzymes

ENZYMES
Enzymes are biological catalysts which do not take part in the chemical reactions but increase the reaction rate.

The chemical properties of proteins enable some of them to function as enzymes which are *biological catalysts* within cells. As with all catalysts, they do not alter the equilibrium of a chemical reaction but they do enhance the *rate* of a reaction. However, they are different to most catalysts in their *specificity*, often only catalysing one reaction. In addition, they operate at relatively low temperatures and pressures compared to industrial inorganic catalysts.

As outlined below, substrates combine with the active site of an enzyme (E) to form an enzyme-substrate complex (ES). Once the substrate has bound, catalysis occurs, and the new products (P) of the reaction are released from the active site which is then free to accept further substrates.

$$\mathbf{E + S \rightleftharpoons ES \longrightarrow E + P}$$

Ex 5.2
What factors could alter enzymatic reactions?

Enzymes are essential to lifeforms. They catalyse all the biochemical reactions that occur within a cell – many hundreds of them which are collectively termed *metabolism.* Some typical enzymes include: Oxidoreductases, which is responsible for oxidation and reduction processes in the cell; Hydrolases which hydrolyse carbohydrates, fats, lipids into smaller molecules; and Lyases which catalyse the addition or removal of substituent groups.

5.3 Metabolism

> Section 5.3 requires a study time of about ½ hour

5.3.1 Chemical Reactions

In Table 5.2, the components of the cell are listed but they are not static. There is a constant turnover of the chemicals in many hundreds, possibly thousands,of enzyme-catalysed chemical reactions. This is metabolism, i.e. all the chemical reactions that occur within a cell or organism.

Metabolism has many different sets of chemical reactions known as metabolic pathways. In order to simplify this complex situation, two major types of pathways can be considered. *Catabolism* is the breakdown of large molecules into small ones to release energy. *Anabolism* is the building up of small molecules to produce large molecules, which requires energy. In other words, catabolism provides the energy to allow anabolism to occur:

> **CATABOLISM**
> Catabolism (*degradation*) breaks large molecules down into smaller ones and releases energy.

CATABOLISM ⇒ ENERGY ⇒ ANABOLISM

Therefore the cell needs some kind of energy transfer 'currency'. In most industrial processes energy is in the form of heat and the transfer currency might be steam. In living things that currency is not heat but the chemical *adenosine triphosphate* or ATP. So, during catabolism ATP is formed from the precursor molecule adenosine diphosphate, inorganic phosphate and the energy released from the breakdown of organic molecules:

> **ANABOLISM**
> Anabolism (biosynthesis) uses energy to build large molecules from small ones.

ADP + Phosphate + Energy = ATP

During anabolism, ATP is broken down to ADP and inorganic phosphate, liberating energy that can then be used in the synthesis of biochemicals:

ATP = ADP + Phosphate + Energy

Catabolism and anabolism are therefore related as shown in Figure 5.2. The breakdown and synthesis of these biochemicals are undertaken, of course, by enzyme mediated reactions.

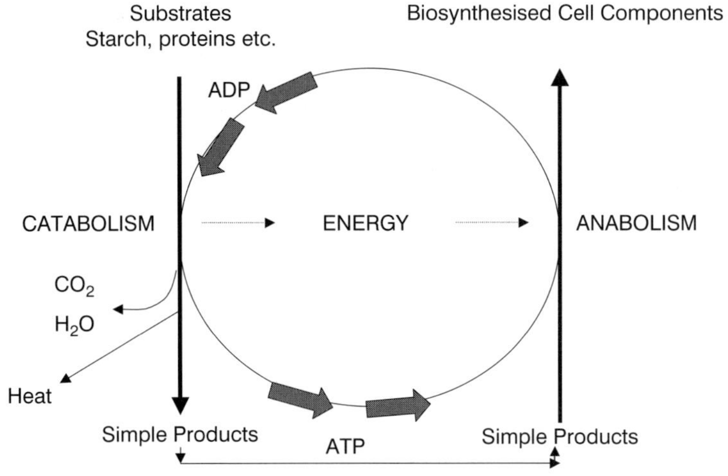

Fig. 5.2 – Catabolism and anabolism

Ex 5.3
How are carbohydrates broken down in aerobic respiration?

5.3.2 Metabolic Pathways

A more detailed study of the reactions involved in catabolism reveals that there are a number of points at which ATP is produced. These are identified in Figure 5.3 for the aerobic respiration of proteins, carbohydrates and lipids.

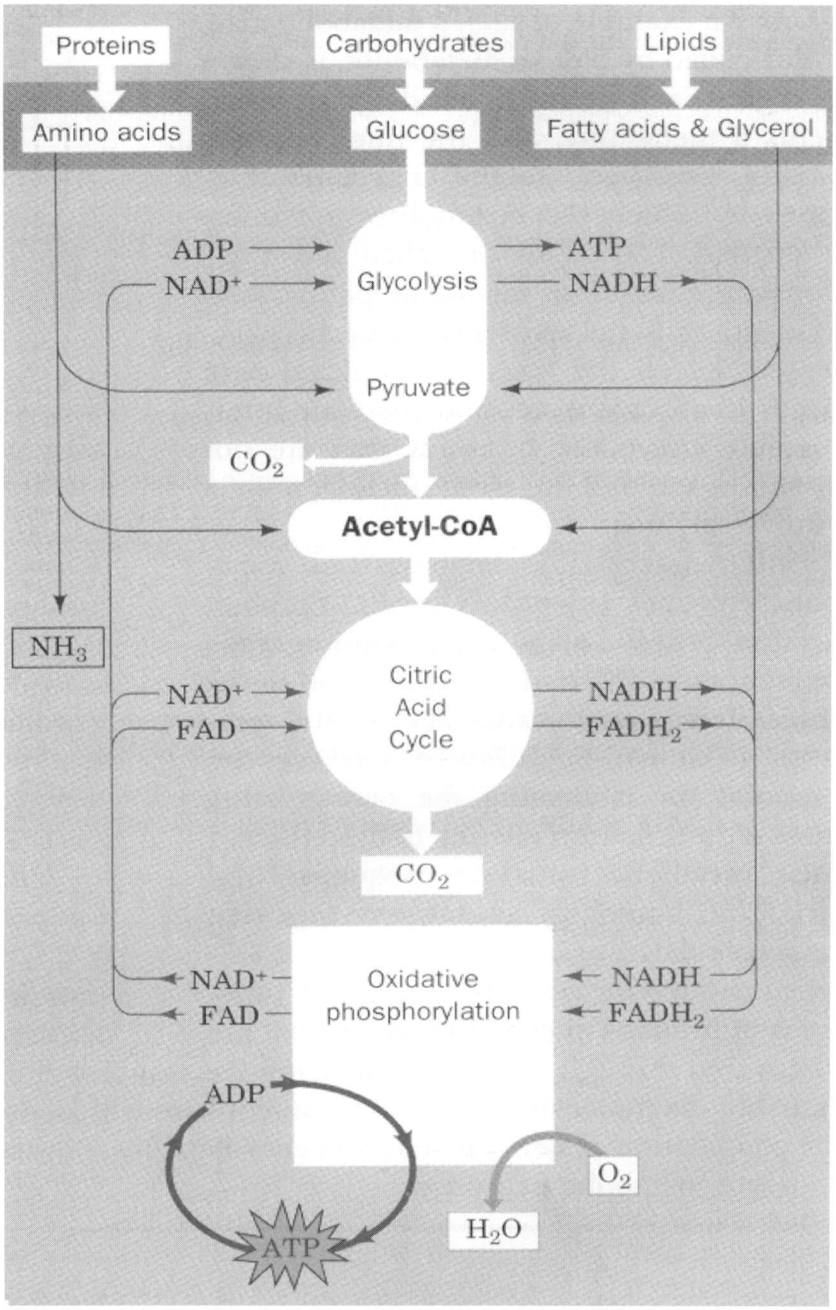

Fig. 5.3 – Respiration pathways for the breakdown of proteins, carbohydrates and lipids

When an organic compound is oxidised/degraded biochemically it loses electrons in the form of hydrogen ions. Similarly an organic compound can receive electrons in the form of hydrogen ions when it is reduced. These hydrogen ions need to be transported during biochemical reactions, often to other parts of the cell, and this is achieved by using co-enzymes like NAD^+/NADH (Nicotinamide Adenine Dinucleotide/Nicotinamide Adenine Dinucleotide Hydrogen) and FAD^+/FADH (Flavin Adenine Dinucleotide and Flavin Adenine Dinucleotide Hydrogen). When organic compounds are oxidised to CO_2, pairs of hydrogen atoms (electrons) are released and transferred to NAD^+. Next these hydrogen atoms are passed through a sequence of electron transport chain reactions, which predominantly occur during oxidative phosphorylation, during which ATP is regenerated from ADP. At the end of this journey these hydrogen atoms are combined with oxygen to produce water. Thus it can be seen that the amount of ATP is related the amount of NADH produced.

Respiration metabolic pathways include aerobic respiration where micro-organisms use oxygen as the ultimate electron acceptor in catabolism. Anaerobic respiration is also possible when certain microbial groups are capable of using oxidised inorganic ions (nitrate, sulphate and carbonate) instead of oxygen as terminal electron acceptors in catabolism. Otherwise, fermentation occurs where organic molecules are broken down into a component that is oxidised and another component that is reduced while the micro-organism derives some energy from the process (Figure 5.4).

The main difference between respiration and fermentation metabolic pathways is the amount of energy that can produced. If one mole of glucose is taken as the starting point in respiration: two moles of ATP are generated through the production of pyruvate; a further two moles of ATP are generated in two turns of the Citric Acid Cycle; finally 24 moles of ATP are generated in the electron transport chain during oxidative phosphorylation. Therefore, 28 moles of ATP can be generated from the oxidation of one mole of glucose. Fermentation on the other hand only yields two moles of ATP for each mole of glucose fermented during the production of pyruvate. This explains why aerobic micro-organisms can grow faster than their fermentation counterparts.

ATP Production
Respiration produces 28 moles of energy yielding ATP whilst fermentation produces only 2 moles for each mole of glucose broken down.

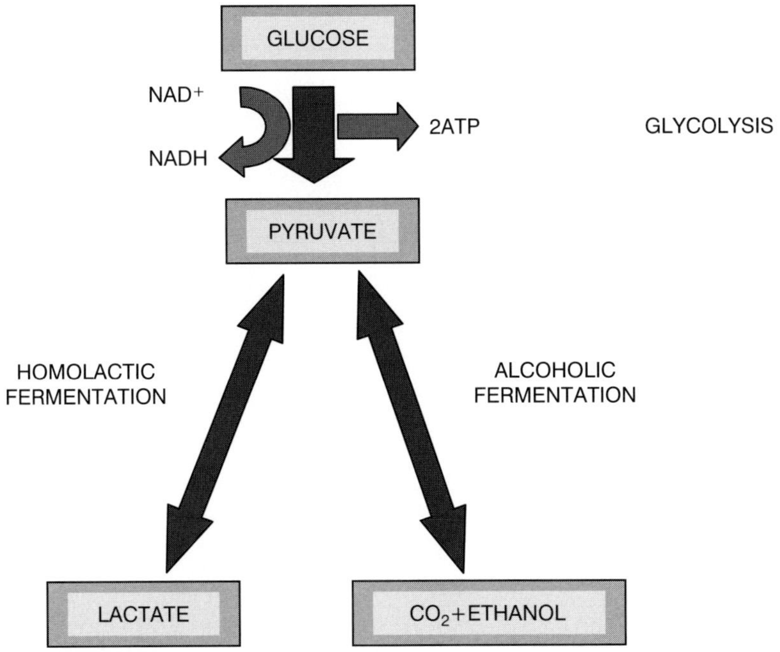

Fig. 5.4 – Fermentative metabolism

5.4 Self Assessment Questions

SAQ 5.1 What category of biochemical macromolecules do the following belong to?

a) Adenine
b) Glycerol
c) Starch
d) DNA
e) Enzymes

SAQ 5.2 What role do the following enzymes have in metabolism?

a) Oxidoreductases
b) Hydrolases
c) Lyases

SAQ 5.3 Which of the following terms can be linked to (a) Catabolism and (b) Anabolism or to both?

a) Biosynthesis
b) Energy
c) Degradation
d) ATP

SAQ 5.4 Name three major types of electron acceptors in metabolic pathways and the normal metabolic end products for carbohydrates being degraded in these metabolic pathways?

5.5 Glossary of Terms

Aerobe	A micro-organism capable of growing in the presence of air.
Aerobic	In the presence of air.
Algology	The study of microscopic plants (algae).
Anaerobe	A micro-organisms capable of growing without air. May be facultative or strict.
ATP	Adenosine tri-phosphate. The energy currency of cells.
Bacteriology	The study of bacteria.
Biochemistry	The study of the chemistry of living things.
Cell	The basic unit of life. All living things are made up of one or more cells.
Chromosome	A structure inside the nucleus that carries the DNA from generation to generation.
Classification	The way things are named.
Culture	Micro-organisms growing in liquid or on solid growth media.
DNA	Deoxyribose nucleic acid. The chemical that is the basis of genetic material. Determines what a cell and hence an organism will look like and how it will function.
Ecology	The study of how living things interact with the environment.
Enteric	Of the gut.
Eucaryote	The more complex of the two basic types of cell e.g., animal cells.
Exponential growth	Growth of a culture where the number of cells doubles over a given period of time.
Extracellular	Outside the cell.
Facultative anaerobe	A micro-organism that can grow with or without air.
Genetics	The study of inheritance.
Growth kinetics	The study of the speed at which bacteria grow.
Growth media	Liquid or solid material which contains nutrients for the growth of micro-organisms.
Immunology	The study of the immune system i.e., how animals (including man) react to infection by micro-organisms.
Intracellular	Inside the cell.
Lipids	Fats.
Logarithmic growth	See exponential growth.
Metabolism	All the chemical reactions taking place inside a cell or organism.
Microbe	Same as micro-organism.
Microbiology	The study of microscopic living things.
Micro-organism	A microscopic living thing.
Mitochondria	The parts of a eucaryotic cell where energy is produced.
Morphology	The study of shape.
Mycology	The study of fungi.
Nucleus	The part of a cell that contains the chromosomes (DNA).
Nucleic acid	DNA, genetic material.
Nutrients	Food.
Organism	A living thing.
Pathogen	Something which causes a disease.
Physiology	The study of the way living things work, or function.
Procaryote	The simpler of the two basic types of cell, e.g., bacteria.

Protein	A large chemical compound that can take many forms and performs many functions within cells.
Protozoology	The study of microscopic animals (protozooans).
Ribosomes	The areas within a cell where proteins are made. They convert the genetic message in DNA into the correct protein.
Saprophyte	An organism which grows on another organism.
Strict anaerobe	A micro-organism that can only grow in the absence of air. In fact, it is killed by air.
Toxin	Poison.
Viable	Capable of living, alive.

5.6 Solutions to Exercises

Ex 5.1 Cell membrane contains lipids and proteins; Cell wall contains polysaccharids and peptides; Chromosome (DNA) contained in cell attached to the cell membrane; Free ribosome (protein and rRNAs) dispersed throughout the cell cytoplasm; Amino acids and simple sugars dispersed throughout the cell cytoplasm.

Ex 5.2 Substrate concentration, pH, temperature, ionic strength, presence of toxicants.

Ex 5.3 Carbohydrates are first broken down into their monomeric units – glucose. During glycolysis glucose is converted to the three carbon sugar pyruyvate with the concomitant generation of 2 ATP molecules. This is further broken down to Acetyl-coenzyme A (a two-carbon acetyl unit linked to a coenzyme A). This intermediate is then further oxidised to CO_2 in the citric acid cycle. The reduced coenzymes NADH and $FADH_2$ that are produced during the citric acid cycle are then oxidised during the oxidative phosphorylation process to produce H_2O and ATP.

6. Microbial Kinetics

Contents

Aims and Objectives

This Unit provides and introduction to microbial kinetics which impacts on all aspects of water and wastewater treatment.

After studying these notes you should be able to:

1. explain in your own words the following terms and concepts:

   specific growth rate
   Monod Equation
   yield coefficient
   hydraulic retention time
   specific oxygen uptake rate

2. carry out calculations for predicting:

   hydraulic retention time
   biomass production

It is not essential for you to memorise any equations or constants.

It is important that you are able to correctly complete all the self assessment questions at the end of these notes.

Essential Prerequisites

Prior to undertaking this lecture it is important that you have completed and understood the following lectures from the *Process Science and Engineering* Module:

Unit 4 Fundamentals of Microbiology
Unit 1 Fundamentals of Water Chemistry
Unit 2 Chemical Kinetics and Equilibria
Unit 7 Fundamentals of Process Engineering
Unit 8 Mass and Heat Balances
Unit 9 Introductory Mass and Heat Transfer

6.1 Introduction

Effective control of any system using biological processes is based on an understanding of the basic principles governing the growth of micro-organisms.

We have already looked at the environmental conditions, which have an important effect on the survival of micro-organisms, like pH, temperature, oxygen, water and nutrient and energy requirements. Now we shall begin to look at the growth of micro-organisms and in particular the kinetics of bacterial growth which is a key parameter in the design of treatment processes which use micro-organisms. For example, to ensure that micro-organisms will grow, they must be allowed to remain in the system long enough to reproduce. The length of time for this depends on the micro-organism growth rate, which is related directly to the rate at which they utilise substrates in the waste material.

Section 6.2 requires a study time of about 1 hour

6.2 Microbial Kinetics

6.2.1 Bacterial Growth

The growth of mixed populations of micro-organisms in submerged culture, freely suspended or immobilised in flocs or films, is very complex to study. As a starting point we will consider the case of a single species of bacterium growing in idealised conditions in a *batch culture* (see Unit 9). The changing conditions of cell and *substrate* (that is the component which the micro-organisms are consuming) concentration during batch growth allow a number of limiting conditions to be identified.

Bacteria reproduce by *binary fission*, that is one cell simply divides into two identical cells. This is called the *cell cycle* or *cell division cycle*. The time taken for this cycle of events is called the *doubling time* and the time taken for this to occur is dependent on the species and the conditions of growth. The doubling time can range from minutes to days, depending on the environmental conditions. Since cell growth leads to cell division, producing a daughter cell, which is also able to grow and divide, a large population of cells can build up quickly under favourable conditions. In liquid medium, this growth can be seen by an increase in the turbidity of the medium.

6.2.2 Phases of Growth

Ex 6.1
What typical conditions could limit micro-organism growth?

In a *batch culture* system (see Unit 9), which contains an initial, limited amount of nutrient, once inoculated with bacteria, the composition of the medium will change continuously as nutrients are used up and waste products accumulate and this will affect the rate of growth of the organism. It is necessary to collect experimental data of the variables of interest before a model can be constructed. Figure 6.1 shows a characteristic curve of bacterial cell number

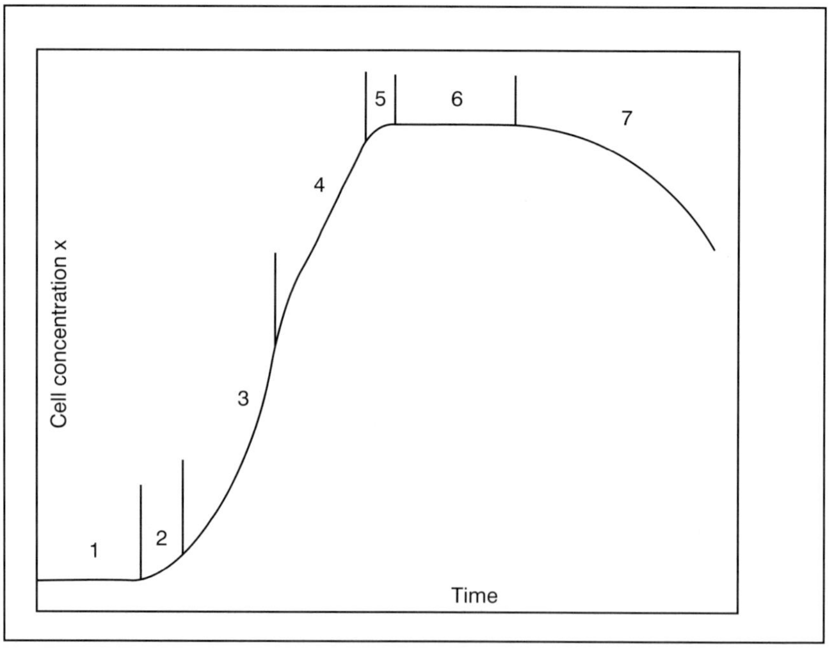

Fig. 6.1 – Phases of growth in batch culture

as a function of time. Note that it is very important to plot raw data in the way that it was obtained before assuming some pre-determined model for the system. Many authors plot log x versus time, wrongly, before the data has been shown to fit a logarithmic function.

The culture will pass through a number of different phases – all of which take place under *unsteady state conditions* (see Unit 8) – and the various phases of growth are identified as follows:

CELL TYPES
Micro-organism growth occurs in seven phases:
Lag
Acceleration
Unlimited Growth
Linear Growth
Deceleration
Stationary
Decline or death
All take place at unsteady state

1. *Lag phase.* Following the inoculation of a culture medium with bacteria, a period is observed in which no measurable change in number occurs. This is attributed to differences in the environmental conditions, e.g., pH, osmolarity or even a change in the carbon source, but it also might be due to bacteria scavenging trace nutrients to inhibitory concentrations within the cells. This phase is often considered as a period of adaptation and it is advantageous to the process if the length of time of the lag phase is minimised.
2. *Acceleration phase.* In this period, the cells start to reproduce and the rate changes quickly up to the maximum possible without nutrient limitation.
3. *Unlimited growth phase.* This is often called the log phase or the exponential growth phase and is of major interest because it is here that most of the cell mass is produced. Cells grow at a rate determined by the inter-cellular enzyme reactions, selecting and transporting nutrients as required by the stoichiometry. The rate of cell growth is the maximum achieved under the culture conditions and the number of cells can be seen to double at a constant rate. To plot a graph of the number of cells against time will produce a curve which will increase sharply, and the simple arithmetic scale used in the graph will soon be inadequate to cope with the increasing number of cells. To over come this problem, a plot of the logarithm of the number of cells against time will produce a linear graph with more manageable data.
4. *Linear growth phase.* In many growth processes there is a period of linear growth. This is brought about by a nutrient limitation of some kind. A common cause of linear growth is the supply of a nutrient at a constant but limiting rate. Oxygen supply limitation is a particularly common cause. However, it can arise from limitations of trace element and vitamin supply.
5. *Deceleration phase.* This phase is very important in many microbial processes because it is the observed effect of some major nutrient limitation on the growth rate. During this phase, different nutrient limitations have quite differing effects on the physiology of the organisms and therefore on products that they might synthesise. The range of physiological conditions experienced during this short phase can be exploited during continuous culture processes.
6. *Stationary phase.* This phase begins when the cells can no longer reproduce. The growth rate depends on a number of factors including the substrate concentration, very high cell concentrations, low partial pressure of oxygen and the accumulation of toxic and metabolic end products. During this phase, storage material may be consumed, but only very sensitive cells die off. As long as there is enough energy to support cell maintenance, then the cells can remain viable for a good length of time.
7. *Decline phase.* Also called the death phase. This phase can be caused by either an increase in the rate of cell death exceeding the rate of formation of new cells or by existing cells using up stored materials such as polyhydroxy butyrate (PHB), glycogen or lipids in a process called endogenous metabolism. In this phase, many of the cells will be undergoing *autolysis*, which is literally a form of self destruction where the cells

burst themselves. Often the cells are intact but non-viable. It is common to see *cryptic growth*, when any components of cells that are released on autolysis can be used by other cells as a source of nutrient.

6.2.3 Specific Growth Rate

In a batch reactor where there is no nutrient limitation or depletion and no loss of cells due to endogenous metabolism or death, growth of biomass will be exponential during the unlimited growth phase (3).

SPECIFIC GROWTH RATE
Specific growth rate, μ, is defined by:

$$x_t = x_0 e^{\mu t}$$

The *specific growth rate* is defined as the quantity of new biomass formed per unit of original biomass per unit of time. The following equation can be used to calculate the specific growth rate:

$$\ln(x_t/x_0) = \mu t \tag{6.1}$$

or

$$x_t = x_0 e^{\mu t} \tag{6.2}$$

where x_t = population at time t, number of cells per ml
x_0 = population at time 0
t = time, h
μ = specific growth rate

Equations 6.1 and 6.2 are analogous to the first order chemical reaction (Equations 2.7 and 2.8 in Unit 2, Section 2.1), representing exponential decay. Therefore the rate constant in a chemical reaction is analogous to the specific growth rate in microbial kinetics.

Ex 6.2
Calculate the maximum specific growth rate and doubling time for a culture of bacteria from the following data:

T (h)	*Count*
1	*4×10^4*
2	*1.6×10^5*
3	*6.3×10^5*
4	*2.5×10^6*
5	*1.0×10^7*
6	*4.0×10^7*

As the culture is growing exponentially, this will be the maximum specific growth rate or μ_m (generally in h^{-1}). For a given set of conditions, this will be a constant which is a function of the species, temperature, pH and nutrient conditions.

The *doubling time* (t_d) is analogous to the half-life for first order decay. It is defined as the time taken for the population to double in size and is related to the specific growth rate. Substituting $x_0/2$ for x_t in Equation 4.1 gives the *Lokta Equation*:

$$t_d = \frac{\ln 2}{\mu_m} = \frac{0.693}{\mu_m} \tag{6.3}$$

Doubling time is useful for estimating biomass growth but has limited value in engineering design. Table 6.1 shows a list of typical values of maximum specific growth rates and doubling times for different organisms.

Table 6.1 – Specific growth rates and doubling times

Organism	μ_m (h^{-1})	T_d (h)
Fungi	0.2	3.5
Yeasts	0.45	1.5
Bacteria	2.0	0.35

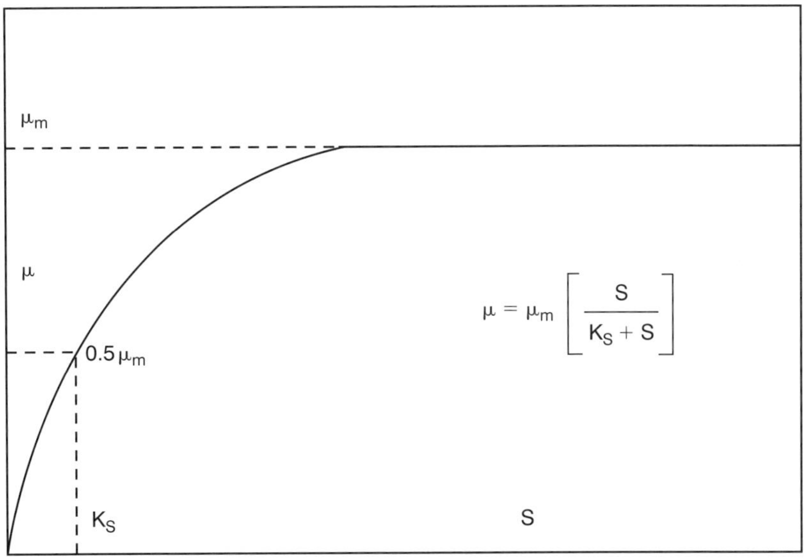

Fig. 6.2 – Nutrient concentration, s vs growth rate, μ

6.2.4 Nutrient Limitation

Nutrient limitation occurs when the growth of the organism is restricted by a single nutrient, for example, carbon, nitrogen or phosphorus, which is termed a limiting nutrient. At this stage in the growth cycle, a change in the growth rate can be observed. The way in which the specific growth rate falls from its maximum value during the deceleration phase as the substrate concentrate becomes limiting, was first reported by Jacques Monod for *Escherichia coli*. The shape of the data is shown in Figure 6.2 and has a similar form to simple enzyme reaction kinetics.

The form of these data can be represented mathematically in a variety of ways but most commonly by the Monod Equation:

$$\mu = \mu_m \frac{s}{K_s + s} \tag{6.4}$$

where s $=$ nutrient concentration
μ $=$ specific growth rate
μ_m $=$ maximum specific growth rate
K_s $=$ saturation coefficient for the growth limiting nutrient

MONOD EQUATION
The Monod Equation relates specific growth rate to nutrient concentration:

$$\mu = \mu_m \frac{s}{K_s + s}$$

The value of K_s can be obtained from Figure 6.2 as the value of s when $\mu = \mu_m/2$. Values of K_s for glucose and glycerol range from 2 to 25 mg/L whilst for oxygen it is 0.05–0.50 mg/L. This parameter is a measure of the ability of the organisms to uptake the particular nutrient under consideration.

If $s \gg K_s$, then the value of $s/(K_s + s)$ approaches unity. In this case, $\mu = \mu_m$ and the specific growth rate is at a maximum and independent of the nutrient concentration.

If $s \ll K_s$ the Monod equation reduces to:

$$\mu = s\frac{\mu_m}{K_s} \tag{6.5}$$

where μ_m/K_s is a constant and the specific growth rate is proportional to the nutrient concentration.

Ex 6.3
What is the maximum specific growth rate if the glucose substrate concentration is 100 mg C l^{-1}. Assume K_s is 5 mg C l^{-1} and the growth rate is 0.2 h^{-1}.

6.2.5 Death Phase

Ex 6.4
If a bacterial population is in decline what will the population be after 2 hours? Assume this population is in a closed culture that has a carrying capacity of 3×10^7 and an initial population of 2.5×10^6 cells per ml with a specific growth rate of $5 \times 10^{-3} min^{-1}$:

The occurrence of autolysis and cryptic growth, together with any endogenous metabolism, ensure that slow cell growth is occurring and that cell numbers never reach zero. However, the cell concentration is seen as a general downward trend.

The Pearl-Verhulst equation can be used to calculate the population size during this part of the growth cycle. This equation models the logistic equation on the carrying capacity of the medium. Here the assumption is made that the growth rate is dependent upon the ratio of the population size and the carrying capacity.

$$x_t = \frac{x_f}{1 + e^{c-\mu t}} \tag{6.6}$$

where x_t = population at time t, number of cells per ml
x_f = carrying capacity
x_0 = population at time 0
c = a constant
μ = specific growth rate
t = time, h

The constant c can be calculated using the following equation:

$$c = \ln\left(\frac{x_f - x_0}{x_0}\right) \tag{6.7}$$

6.2.6 Diauxic Growth

DIAUXIC GROWTH
Diauxic growth describes the situation where one nutrient in a mixture is preferred to another.

If a bacterium growing in a liquid medium is provided with a mixture of two different nutrients, it may use one in preference to the other. The second substrate is only used after the first one has been completely exhausted. For example, if *E. coli* is provided with both glucose and sorbitol, glucose will be used in preference to sorbitol. Sorbitol will only be used when all the glucose has been used. During the transition from glucose utilisation to sorbitol utilisation, growth may slow down or even stop and the resulting growth curve has a distinct pattern of the type shown in Figure 6.3.

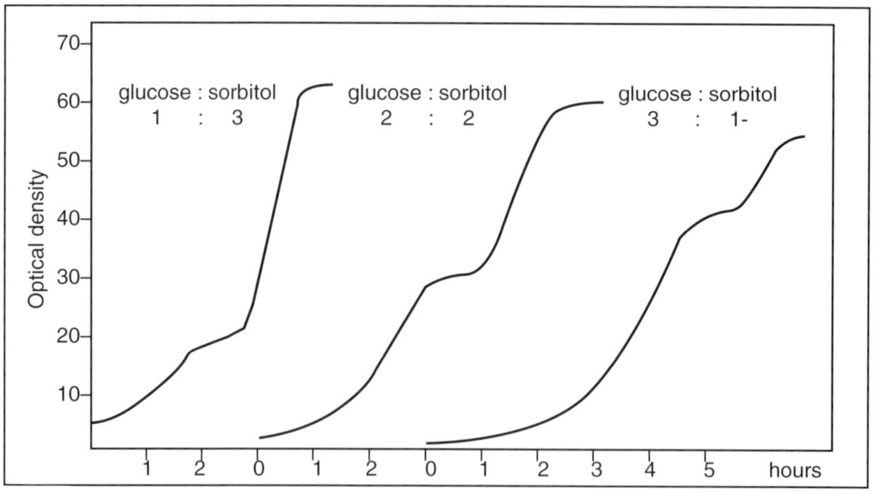

Fig. 6.3 – Pattern of diauxic growth

6.3 Continuous Cultures

Section 6.3 requires a study time of about 1 hour

6.3.1 Production of Biomass

Our analysis of bacterial growth in fixed volume batches above is of only limited value in engineering terms. Most of the biological treatment processes used in water and wastewater treatment are continuous flow processes, that is a *biomass* is retained inside a reactor and the stream to be treated is fed continuously to the reactor and treated product is withdrawn at the same rate. This continuous culture allows the cells to be grown under constant and controlled conditions so that the models of batch growth do not apply.

CONTINUOUS CULTURES
The simplest form of continuous reactor is the completely mixed stirred tank or "CSTR"

The classical activated sludge reactor is a continuous stirred tank (CSTR) type (see Unit 10) and constant agitation is required to ensure a rapid mixing of influent with the growing cells. Continuous cultures have two main advantages over batch culture. Firstly, the cells are provided with a consistent environment, where fluctuations in the concentrations of nutrients and waste products are kept to a minimum. Secondly, by controlling the rate of flow of the medium through the reactor, the rate of bacterial growth can also be controlled. If required, growth can be kept at the exponential rate.

The growth kinetics are different in a continuous culture from those of a batch culture. It is assumed that there is a single stage completely mixed bioreactor fed with nutrients dissolved in water as shown schematically in Figure 6.4. One component will be in a limiting concentration compared with the others and so will determine the yield and physiological condition of the cells.

The assumptions for the CSTR reactor model are:

(a) Constant liquid volume and a well-mixed system, so that the outlet stream concentrations are the same as the values in the vessel.
(b) No cell death or endogenous metabolism.

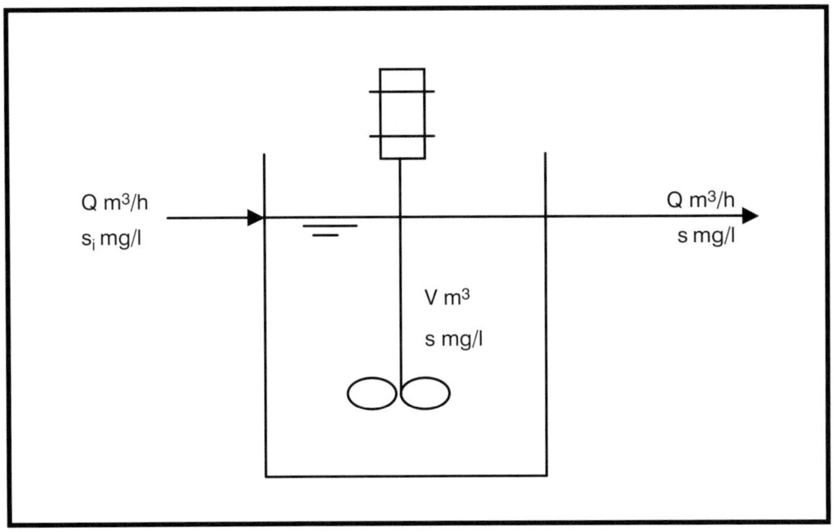

Fig. 6.4 – Continuous culture system

6.3.2 Mass Balance on Biomass

HYDRAULIC RETENTION
The residence time or hydraulic retention time of a flow of $Q\,m^3.h^{-1}$ in a reactor of volume $V\,m^3$ is given by

$$\tau = V/Q$$

Although fresh nutrients are added continually, all other factors will remain constant. We can define, the *mean residence time*, τ, or *hydraulic retention time* (HRT) as:

$$\tau = \frac{V}{Q} \tag{6.8}$$

where V = reactor volume, m^3
Q = feed flow rate $m^3.h^{-1}$

HRT has units of h and it is sometimes convenient to refer to the *dilution rate*, D which is defined as the reciprocal of the HRT and has units h^{-1}.

It is important to remember that the hydraulic retention time or dilution rate of a continuous culture system determines the specific growth rate of the organisms. Also, to avoid wash out of the organisms from the system, the rate of inflow of the medium must not exceed the rate of growth.

YIELD COEFFICIENT
The yield coefficient Y is the mass (kg) of cell material produced by the metabolism of 1 kg of a specified nutrient.

The mass of cells, m, produced in the reactor is given by

$$\mathbf{m = YQ(s_i - s)} \tag{6.9}$$

where m = biomass production, kg/h
s_i = concentration of limiting nutrient in the influent, $kg.m^{-3}$
s = concentration of limiting nutrient in the reactor, $kg.m^{-3}$
Y = yield coefficient, kg biomass per kg nutrient

Typically, $s \ll s_i$ so that Equation 6.10 can be approximated as:

$$\mathbf{m = YQs_i} \tag{6.10}$$

or

$$\frac{\mathbf{m}}{\mathbf{V}} = \mathbf{Y}\frac{\mathbf{s}}{\tau} \tag{6.11}$$

where m/V is the rate of biomass production per unit volume of reactor, or *productivity* in $kg.m^{-3}.h^{-1}$. Now **Ys** is the concentration of biomass in the reactor produced from **s** $kg.m^{-3}$ of nutrient. If the biomass concentration is written as $\mathbf{c_B}$ we have:

$$\frac{\mathbf{m}}{\mathbf{V}} = \frac{\mathbf{C_B}}{\tau} \tag{6.12}$$

6.3.3 Oxygen Limitation

In order to achieve a high productivity from an aerobic continuous culture, like the *activated sludge process*, it is necessary to run at a high dilution rate or low hydraulic retention time (limited by μ_m) and a high value of m. From Equation 6.9, a high value of m will be achieved from a high value of s_i. However, as m is increased there will be a maximum value at which oxygen supply will become limiting.

For aerobic organisms a specific oxygen uptake rate ($\mathbf{q_0}$) can be determined with the units of kg O_2 $(kg\,cell)^{-1}$, h^{-1}, and the rate at which oxygen is transferred into the water must exceed this minimum.

In Unit 9 we derive the equation for oxygen transfer into solution as:

$$\mathbf{n = k_La(c^* - c)} \qquad \mathbf{(6.13)}$$

where n $= $ rate of mass transfer, $kg.h^{-1}$
k_La = absorption coefficient, $m^3.h^{-1}$
c_{DO} = dissolved oxygen concentration in influent, $mg.l^{-1}$
$c_{DO}{}^*$ = equilibrium dissolved oxygen concentration, $mg.l^{-1}$

A maximum cell concentration, m_{max}/V, can be determined for a given absorption coefficient and permissible minimum dissolved oxygen concentration, $c_{DO,min}$:

$$\frac{m_{max}}{V} = \frac{k_La(c^*_{DO} - c_{DO,min})}{q_0\tau} = \frac{c_B}{\tau} \qquad \mathbf{(6.14)}$$

6.4 Self Assessment Questions

SAQ 6.1 An exponentially growing culture increases from 1.5×10^5 cells per ml to 5.0×10^8 cells per ml in 6 hours. Calculate: The specific growth rate; and The doubling time.

SAQ 6.2 After a day the number of cells per ml for an unlimited exponentially-growing culture is 3×10^{18} cells per ml. If the doubling time is 0.5 h what was the initial population?

SAQ 6.3 Calculate the HRT at which a continuous flow biomass unit must operate in order to achieve a maximum cell production of 28 kg cells.$m^{-3}.h^{-1}$, within the limit of oxygen supply.

Given:
$q_0 = 8 \times 10^{-2}$ kg O_2 (kg cells)$^{-1}.h^{-1}$
$k_La = 250\,h^{-1}$
$c^*_{DO} = 6 \times 10^{-3}\,kg.m^{-3}$
$c_{DO,min} = 1 \times 10^{-3}\,kg.m^{-3}$.

6.5 Nomenclature

c_B = biomass concentration, $kg.m^{-3}$
c_{DO} = dissolved oxygen concentration, $kg.m^{-3}$
c_{min} = minimum permissible oxygen concentration, $kg.m^{-3}$
k_La = absorption coefficient (h^{-1})
q_0 = specific oxygen uptake rate (kg O_2 (kg cells)$^{-1}$.h^{-1})
s = nutrient concentration, kg/m^3
si = influent nutrient concentration (kg/m^3)
t = time, h
t_d = culture doubling time, h
x = population, $cfu.ml^{-1}$
x_f = carrying capacity, $cfu.ml^{-1}$

D = dilution rate (h^{-1})
Q = feed flow rate $m^3.h^{-1}$
K_s = saturation coefficient for the growth limiting substrate
V = volume, m^3
Y = yield coefficient, $kg.kg^{-1}$

μ = specific growth rate, h^{-1}
μ_m = maximum specific growth rate, h^{-1}
τ = hydraulic residence time, h

6.6 Solutions to Exercises

Ex 6.1 Low $< 5°C$ or high $> 70°C$ temperature; Lack of O, C, N, P and S nutrients, monoatomic ions Na, K, Mg, Ca and Cl and trace elements; Alkaline pH conditions; High salinity levels; Lack of water

Ex 6.2 This can be done in one of two ways:

(a) Plot the data as $\ln x$ versus t and determine the slope of the straight part of the line.

(b) $\mu_m = (1/t) \times \ln(x_t/x_0)$

$= (1/(5 - 1)) \times \ln(1.0 \times 10^7/4 \times 10^7) = 1.38\,h^{-1}$

$t_d = 0.693/\mu_m = 0.693/1.38 =$ **0.5 h**

Ex 6.3 $0.2\,h^{-1}$ as $s \gg Ks$ so $\mu = \mu_m$

Ex 6.4 $c = \ln\,(x_f - x_0)/x_0$

$c = \ln\,(3 \times 10^7 - 2.5 \times 10^6)/2.5 \times 10^6$

$c = \ln 11 = 2.40$

$x_t = x_f/1 + e^{c-\mu t}$

$x_t = 3 \times 10^7/1 + e^{2.4 - 0.3 \times 2}$

$x_t = 3 \times 10^7/1 + 6.05$

$\mathbf{x_t}$ **= 4255531.95 cells per ml (4.2 × 10^6 cells per ml)**

7. Fundamentals of Process Engineering

Contents

Aims and Objectives

This unit is an introduction to the concepts of process engineering. The concepts will be explored further in the following units of the *Process Science and Engineering* Module:

Unit	Topic
8	Mass and Heat Balances
9	Introductory Mass and Heat Transfer
10	Reactor Design Theory
11	Engineering Hydraulics
12	Particle Settlement

After studying this Unit you should be able to:

1. explain in your own words the following terms and concepts:
   - dimensions
   - dimensional analysis
   - unit operations

2. carry out calculations for:
   - conversion of measurement units

3. draw and understand
   - a basic block diagram
   - a Process Flow Diagram

4. interpret
   - a Piping and Instrumentation Diagram

It is not essential for you to memorise any of the constants or equations given in this Unit.

It is important that you are able to correctly complete all the self assessment questions at the end of these notes.

Essential Prerequisites

It is not necessary to have completed any other Units before undertaking this Unit.

7.1 Introduction

ENGINEERING ECONOMICS
Economics is a fundamental to engineering decision making.

Engineering is the economic application of scientific principles to benefit society. Why *economic*? Because engineers always strive to find ways of achieving the same end result for a lower cost by, for example, developing better materials of construction, machines which utilise energy more efficiently or manufacturing techniques which use less labour. *Process Engineering* is that branch of engineering which deals with the design, construction and operation of process plant in which matter undergoes a change in state or chemical composition. This covers many different industries including oil refining, pharmaceuticals manufacture, food and drink, brewing, power generation, chemicals production and many more. The largest bulk chemical industry is water supply. In the UK alone, the water supply industry delivers to points of use over 17 million tonnes of a product per year which is 99.95% pure.

The great engineers of the past like de Lesseps, Brunel, Westinghouse and Mond, worked largely autonomously but, these days, engineering projects draw on expertise from many different disciplines – a knowledge base beyond any single person's capability. The design and construction of a water works will involve process engineers, who are responsible for the design of the chemical treatment process; mechanical engineers, who select materials and specify rotating machinery like pumps; electrical engineers who provide the power for pumps and other equipment; civil engineers who are responsible for foundations, buildings and, often, for process reactors; and controls engineers, who will provide automatic computer-based controls for the complete works. Other disciplines which will provide specialist inputs include chemists, biologists, architects, planners, estimators, quantity surveyors and specialist equipment manufacturers.

The *Project Manager* is the key person who co-ordinates the activities of the team members and ensures that everything happens at the right time. This involves *communication* between the various team members and it is essential that formal procedures are used so that all those involved know about any changes in the design. Engineers communicate in words, drawings and calculations. Although communication is such an important facet of engineering, communication skills are rarely taught formally. It is important to get into the habit of writing clear, concise, logically structured reports and to present calculations in such a way that each step can be followed and checked. Calculations should reflect the accuracy of the source data which are used and measurements should reflect the precision of the measuring device.

Section 7.2 requires a study time of about 1 hour

7.2 Dimensions and Units

7.2.1 Dimensional Analysis

The concept of *dimensions* will be familiar from physics where all measurements can be described by reference to the four basic dimensions of mass (M), length (L), time (T) and temperature (Θ), and all engineering measurements are based on these.

A simple example reveals that area (length multiplied by width) has dimensions of $L \times L = L^2$, and similarly volume has dimensions of L^3. Velocity, which is the distance travelled divided by the time taken, has dimensions of L/T. It is axiomatic that any equation must be dimensionally consistent, that is the dimensions on each side of the equation must be the same. This is not always obvious but can be checked by observation. For example, Stokes' Law, which describes the settling velocity of a particle as follows:

$$v_t = g\frac{(\rho_s - \rho)d^2}{18\mu} \tag{7.1}$$

On the right hand side of the equation:

g	acceleration	$L.T^{-2}$
ρ	density	$M.L^{-3}$
d	diameter	L
μ	viscosity	$M.L^{-1}.T^{-1}$

Ex 7.1
Check the following equation for dimensional consistency and identify any missing terms:

$$v = Q/D^2$$

where v = velocity
Q = mass flow rate
D = pipe diameter

So the dimensional analysis of the right hand side is:

$$\frac{L}{T^2} \times \frac{M}{L^3} \times L^2 \times \frac{L.T}{M} = \frac{L}{T} \tag{7.2}$$

and on the left hand side:

$$v_t \qquad \text{velocity} \qquad L.T^{-1} \tag{7.3}$$

so the equation is dimensionally consistent.

7.2.2 Mass, Weight and Force

MASS AND WEIGHT
Mass is a measure of quantity.
Weight is a measure of force.
Force = Mass × Acceleration

One point that needs to be carefully noted is the difference between *mass, weight* and *force* both of which we speak of in terms of the same units (typically kg or lb). Mass is volume multiplied by density and force is mass multiplied by acceleration. So the force, F, with which the earth attracts a mass of m kg is:

$$F = m \times g \tag{7.4}$$

This is the *weight* of the mass m kg. Because everything is weighed by comparing this force with the force with which the Earth attracts a known mass and, because g is a constant (9.81 $m.s^{-2}$), masses can be equated. This would not be the case were the weight of a mass on Earth to be compared that of the same mass on Mars where the gravitational acceleration is different: this would be a comparison strictly of *forces*. Care should therefore be taken when discussing kg weight, kg mass and kg force: mass has dimensions M whilst weight has dimensions of force, that is MLT^{-2}.

Ex 7.2
Sludge produced from a sewage treatment works has a density of 1020 kg/m^3. If the volume flow is 50 mg^3/day what is the mass flow in kg/h?

In the Imperial system of measurement, force is measured in *poundals* (1 poundal = 32.2 pounds force); in the metric cgs system the unit is the *dyne* (1 dyne = 981 grams force) and in the SI system, which is normally used, it is the Newton (1 Newton = 9.81 kilogram force).

In many applications, pressure is important (Unit 11). Pressure P is defined as force, F, divided by the area over which it acts, that is:

$$P = F/A \qquad (7.5)$$

PRESSURE
Pressure is force divided by the area over which it acts.
Pressure = Force/Area

Pressure has dimensions of $ML^{-1}T^{-2}$. As an example, the mass of a column of water $2\,m^2$ in area and 12 m deep would be $12 \times 2 \times 1000 = 24000$ kg. The downward force exerted by it would be $24000 \times 9.81 = 235440$ N. Dividing by the area gives a pressure of $235440/2 = 117720\,N.m^{-2}$. The $N.m^{-2}$ is also called *Pascal* (Pa), so the pressure would be more commonly written as 117 kPa.

7.2.3 Work, Energy and Power

In order to move a mass, m, through a distance, l, a force, F, must be exerted. The product of the force and the distance through which it is moved is called the *work done* and this has dimensions of ML^2T^{-2}. By doing this work the mass attains *energy*, E. In a 100% efficient, frictionless world the work performed and the energy imparted to the mass would be the same and both would be calculated by:

WORK
Work is done when a force moves through a distance.
Work = Force × Distance

$$E = F \times l = m \times g \times l \qquad (7.6)$$

Although water processing largely involves only mechanical energy, there are other forms of energy such as electrical and thermal. All are measured, in the SI system, in calories, Joules, metre kilogram force (m.kgf) or kilowatt hours (kWh). 1 Joule = 1.0 N.m = 0.102 mg.kg. In the Imperial system a variety of units including foot-poundals, foot-pounds weight and British Thermal Units (BTU).

Ex 7.3
What is the work done in Joules when a mass of water weighing 2 tonnes is raised through a height of 7.5 m?

The *rate* at which work is done is called power and is measured in *Watts*. That is:

$$\text{Power} = \text{Energy/time} \qquad (7.7)$$

POWER
Power is the rate at which work is done.
Power = Energy/Time

In many engineering applications, power is provided by electricity so it is useful to calculate energy and power in units that can be easily converted in electrical power, namely Watts. A Watt is defined as a Joule per second and a Joule is the work done by a force of 1 kg moving through a distance of 1 m. Thus $1\,W = 1\,N.m.s^{-1}$.

Put another way, if an electric motor rated at 1 kW operates for 1 hour, the work done by the motor is 1 kWh. However, the electricity used will be greater than this because no motor can be made 100% efficient; there is always some thermal and mechanical energy loss.

Ex 7.4
Calculate the power needed in kW to raise $50\,m^3/h$ of water to a height of 5 m. The density of water is $1000\,kg/m^3$

7.2.4 Dimensionless Groups

It is often expedient to define quantities that appear frequently in process design calculations and which, when dimensionally analysed as above, have no dimensions. The most common example of such a *dimensionless group* is the Reynolds number which is very useful in defining the nature of flow in pipes. Some of the more common dimensionless groups used in water treatment process engineering are listed in Table 7.1, and are defined subsequently (Unit 9, 9.2.7).

Table 7.1 – Dimensionless numbers (see also Unit 9, 9.2.7)

Name	Symbol	Formula	Application
Archimedes	Ar	$gL^3(\rho_p - \rho)\rho/\mu^2$	Particle settling
Froude	Fr	v^2/gd	Flocculation
Peclet	Pe	dv/D	Mixing and mass transfer
Prandtl	Pr	$C\mu/k$	Mixing and mass transfer
Reynolds	Re	$Dv\rho/\mu$	Flow
Schmidt	Sc	$\mu/\rho D$	Mass transfer
Sherwood	Sh	k/D	Mass transfer
Weber	We	$\rho v^2 d/\sigma$	Bubble drop formation

7.2.5 Units of Measurement

FORMAT FOR UNITS
Scientists generally use the format

$kg.m^{-3}$ and $m.kg.s^{-1}$

Engineers generally use the format

kg/m^3 and m.kg/s

Both are equally acceptable and will be used interchangeably throughout the course.

Most European engineers work in SI units. In the USA the *fps* (foot, pound, second) system is still in common use and filtration rates are still given in gallons per minute per square foot, concentrations in grains per gallon and pressures in pounds per square inch (*psi*). Note that a US gallon is different from the British *Imperial* gallon (1 US gal = 0.83 imp gal). Even in Europe many of the SI units are either too small or too large for convenience and, for example, pressures are commonly measured in bar rather than Pascals, and in the water industry, flows are normally quoted in mega-litres per day (*Mld*) which is the same as thousands of cubic metres per day (*tcmd*).

It is important when dealing with pressure to know whether the pressure is *absolute* or *gauge*. Gauge pressures are measured above atmospheric pressure so

absolute pressure = gauge pressure plus atmospheric pressure

Thus 1 barg = 2 bara and 1 psig = 15.7 psia. This is discussed further in Unit 11, 2.1.

PRESSURE MEASUREMENT
It is essential to be aware of the difference between *gauge* (g) and *absolute* (abs) pressure measurement.

$$P_{abs} = P_g + P_{atm}$$

Concentrations are habitually quoted in parts per million which is the same as mg/kg. In the case of contaminants in water, this is almost identical to mg/l – certainly within the limits of analytical error. However, when dealing with sludges or with concentrated chemical solutions, which have densities significantly different from 1000 $kg.m^{-3}$, this is not the case.

Engineering handbooks will usually provide conversion factors for units such as lb/ft^3 to $kg.m^{-3}$ and even from derived units such as Poise to $kg.m^{-1}s^{-1}$ representing the units of viscosity (Unit 11, 11.2.2).

Most conversions can be calculated using a few factors and a bit of clear thinking. For example, suppose an American reference quotes a filtration rate as 5.5 gpm/ft^2 and this must be converted to an approach velocity in $m.h^{-1}$. Given that:

1 gal = 1.2 US gal
$1\ m^3$ = 220 gal
1 m = 3.28 ft

Ex 7.5
Given that:

1 kg = 2.2 lb
1 inch = 25.4 mm

convert an absolute pressure of 50 psi into kg/cm^2.

the conversion proceeds thus (see table opposite):

$$\frac{\textbf{USgal}}{\textbf{min.ft}^2} \times \frac{\textbf{Impgal}}{\textbf{USgal}} \times \frac{\textbf{m}^3}{\textbf{Impgal}} \times \frac{\textbf{min}}{\textbf{h}} \times \left(\frac{\textbf{ft}}{\textbf{m}}\right)^2 = \frac{\textbf{m}}{\textbf{h}}$$

$$\mathbf{5.5 \times \frac{1}{1.20} \times \frac{1}{220} \times 60 \times (3.28)^2 = 13.45}$$

7.2.6 Accuracy and Precision

When carrying out any sort of calculation it is important to understand the accuracy and precision of the quantities on which the calculation is based. Suppose three shots are fired at a target. Precision reveals how close the shots are to each other (that is how consistently one shoots) and accuracy tells us how close the shots are to the bulls-eye (Figure 7.1).

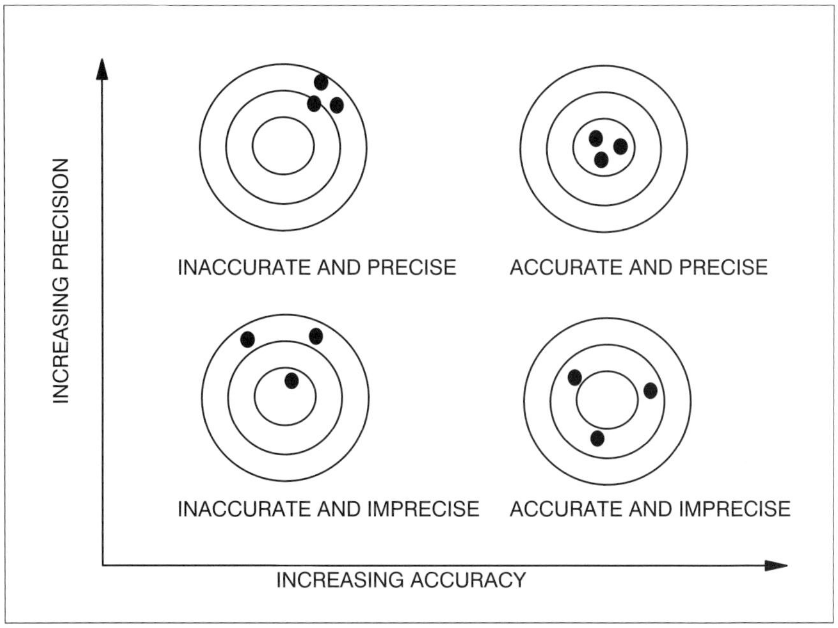

Fig. 7.1 – Accuracy and precision

Suppose the thickness of a sheet of steel has to be measured, and it is known to have been manufactured to a *nominal* thickness of 6 mm. Measuring this thickness with a rule calibrated in millimetres which means that the measurement to the nearest 0.5 mm, for someone with reasonable eyesight. If a micrometer is used the same measurement can be made to the nearest 0.05 mm. The 0.5 mm or 0.05 mm represents the precision of our measurement. If the thickness is measured in various different places on the sheet it will vary. There will usually be a specified *tolerance* which is acceptable. Typically this would be expressed as "6 mm ± 0.5 mm" which means that any thickness between a minimum of 5.5 mm and a maximum of 6.5 mm is acceptable. When the measurement is carried out at several places on the sheet the minimum thickness may be 5.7 mm and the maximum 6.45 mm for a sheet *within tolerance*. But we cannot specify a tolerance of better than 0.5 mm if we only have a mm rule is used to measure the thickness. And, in any event, the precision with which the rule was is made may be poor.

CONVERSION FACTORS
The following are worth memorising

1 inch	= 25.4 mm
1 bar	= 14.7 psi
1 bar	= 10.197 m water
1 bar	= 100 kPa
1 m^3	= 220 imperial gallons
1 imp gal	= 1.2 US gal
1 US gal	= 3.78 litre
1 m	= 3.28 ft
1 mg/l	= 1 g/m^3
1 ha	= 2.47 acre

ACCURACY AND PRECISION
Inappropriate precision is indicative of mathematical naivety.

Ex 7.6
Calculate to six decimal places the velocity (flow/cross sectional area) in m/s for a flow of 75 m³/h of water in the standard "4 inch diameter" pipes whose internal diameters are given in the table below:

PIPE TYPE	*ID, mm*
4" ID	*101.6*
4" NB PVCU Class C	*100.0*
4" NB PVCU Class E	*93.1*
110 mm OD PVCU	*93.6*
110 mm OD Hep$_3$O – max	*104.2*
110 mm OD Hep$_3$O – min	*102.8*
Stanton DI Class K9	*87.8*

When writing down a number a degree of precision is implied. A concentration expressed as “2 mg/l” implies an actual concentration somewhere between 1.5 mg/l and 2.5 mg/l. Similarly, “2.1 mg/l” really means “2.1 ± 0.05 mg/l” and so on. Computers and calculators provide very high levels of precision, but these have no meaning if the accuracy of the input data is questionable; only the user can provide that important factor and truncate or round figures accordingly.

Assumptions are often made about measurements. If a standard 4” NB (*nominal bore*) pipe – most pipes are still made in Imperial (inch) sizes even though we call them the equivalent measurement in mm – has a flow of 75 m^3/h of water through it the velocity of flow (the volume flow divided by the internal cross sectional area of the pipe) can be calculated as being around 2.5 m/s. However, pipes of different specifications vary in internal diameter by a significant amount as the exercise. The whole thing is further complicated because flow meters have a precision of, at best, ±1% so the flow could be in the range 74.3–75.7 m^3/h. All these limitations must be taken into consideration when calculations are carried out to more than about three significant figures.

7.3 Elements of Process Engineering

Section 7.3 requires a study time of about 30 minutes

7.3.1 Tanks

Most process engineering applications, particularly those concerned with water treatment, involve storing liquids and moving them around from one place to another. Within water treatment plants, water is stored in tanks, usually of concrete, suitably protected steel, or plastic construction depending on the size. Wastewater and sludges are often held in lagoons constructed by excavating a hole in the ground and lining it to make it water-tight.

Chemical solutions are usually delivered by road tanker, in 1 tonne *intermediate bulk containers* (IBCs) or in 200, 50 or 25 litre drums. Tanker loads will normally be stored in *bulk storage tanks,* which may be fabricated from steel, glass fibre reinforced plastic (GRP) or polypropylene.

Solid chemicals are usually supplied in bags for easy handling, although some, like lime, may be delivered by road tanker. The chemicals are usually dissolved in water and the solutions stored in plastic tanks. Typically these tanks will hold sufficient for a day's use and are called *day tanks.*

7.3.2 Pumps and Pipework

PUMPS
Pumps are the means of moving water and chemical solutions around the works.

Transferring liquids from one place to another uses pumps of which there are several different types – centrifugal, positive displacement, screw and so on – depending on the type of liquid being pumped, the quantity being pumped and the accuracy with which the pumping has to be carried out. For most duties, centrifugal pumps are the preferred type because they are low cost, efficient and available in a wide range of sizes. Positive displacement *metering* or *dosing* pumps are used for the accurate delivery of chemical solutions at controlled rates. They have a reciprocating plunger or diaphragm which delivers a consistent volume of solution on each stroke.

Water is generally contained in either pipes manufactured from steel, iron, plastic or ceramic or else in open channels of concrete. Pipe sizes range from around 10 mm internal diameter (*nominal bore*) up to 2000 mm for very large flows in water distribution systems.

7.3.3 Valves

VALVES
Valves are used for on/off control of liquid flow, for isolation of equipment and for modulating liquid flow rates.

Valves are used to control how liquids flow through pipes, much like points on a railway track. They can also be used to control or *throttle* the rate of flow. There is a wide variety of different types of valves: butterfly, plug, gate, globe, diaphragm and so on. Where a *tight shut off* is necessary (that is a valve which will not leak) gate valves are the norm but butterfly, plug and diaphragm valves are also used. Diaphragm valves are expensive, but useful where corrosion is likely because the moving parts of the valve do not come into contact with the liquid. For flow control duties, globe valves give the most consistent performance but are very expensive so plug valves are generally preferred.

Valves which are opened and closed by hand using handwheels or levers are referred to as *manual* valves, but valves can also be operated automatically by means of a mechanical device driven by either compressed air (*pneumatic actuator*), an electric motor (*electric actuator*) or an electromagnet (*solenoid valve*). Most actuators either open or close the valve. Sometimes a spring is used to open or close the valve with air pressure moving it to the opposite position. In the event of air failure the spring will automatically open or close the valve so these actuators are referred to as *fail-safe open* or *fail-safe closed* actuators.

Some actuators, often referred to as *positioners*, will *modulate* the valve, that is they will hold the valve at some intermediate position between open and closed controlled by a *control signal*. Automatic control valves of this type are critical to process engineering in applications as diverse as controlling temperature in a heat exchanger, controlling level in a tank, controlling flow for filter cleaning or controlling dissolved oxygen levels in activated sludge plants.

7.3.4 Unit Operations

Water treatment is a process based industry, that is the raw material, *raw water or wastewater,* is subjected to a series (*procession*) of chemical, physical and biological treatment stages each of which will change the characteristics of the water before passing it on to the next stage. These treatment stages are called *unit operations* and the combined effect of the *process* is to convert the defined raw water into the desired quality of product or *treated water*.

Consider the treatment of domestic sewage to reduce BOD prior to discharge. The first stage of treatment in a conventional works will be the removal of suspended solid matter and will use several *liquid-solid separation* unit operations like screening and sedimentation. The second stage of treatment will normally use a *biological reactor* to oxidise dissolved carbonaceous matter which the primary treatment will not remove. However this secondary treatment will generate suspended solids and there will be a second liquid-solid separation unit operation (usually sedimentation) after the reactor. Depending on the quality of treated sewage required, there may be a third treatment stage to remove phosphate by chemical coagulation in a *chemical reactor* and final solids removal by a third liquid-solid separation unit operation (typically filtration).

There are, of course, other ways of achieving the same end result. The raw sewage could be distilled or oxidised chemically using chlorine or ozone, or subjected to an advanced oxidation technique such as supercritical water oxidation. All of these process routes or *flowsheets* would work but may give undesirable by-products or cause harm to the environment or simply be too expensive. We shall discuss this further.

SOLID LIQUID SEPARATION
Liquid-solid separation processes are the most important and widely used unit processes in water treatment. They include screens, clarifiers and filters.

In addition, there will be alternative techniques for each unit operation. For example, there are many different types of liquid-solid separation unit operations – sedimentation, flotation, centrifugation, filter pressing, membrane filtration, deep bed filtration and so on – and these will all have to be assessed before a final choice is made.

In process engineering, as in all engineering disciplines, there are no right and wrong answers; there are different answers some of which will be better than others. The task of the engineer is to assess the alternatives technically, economically and environmentally, to arrive at an optimum design which will satisfy the many constraints.

The unit operations most widely used in water treatment are those which separate suspended solids from liquids. They include gravity separation processes like sedimentation and a range of filtration processes ranging from coarse screening to sub-micron membranes and deep beds of granular materials. Liquid-solid separation unit operations will be feature of almost every module in this course.

Sorption occurs when there is a *concentration gradient*, that is a difference in the concentration of a species, between two phases. Many water treatment unit operations are dependent on sorption processes.

SORPTION
Sorption is a key unit operation in removing dissolved species from water and in dissolving species in water.

Any vessel in which a chemical or biochemical reaction takes place is, in process engineering terms, a *reactor*. In water treatment, reactors may be as simple as a tank with a stirrer, in which chemicals are added to water and mixed, or as complex as a mixed bed ion exchange unit, which undergoes a complicated sequence of chemical reactions during use and regeneration. Reactor design theory is dealt with in Unit 10 of this module and, in most of the other modules in the series, the practical design of reactors is considered in more detail.

7.3.5 Control Systems

Process plants are controlled by *control loops*. Automatic controllers simply take the place of a human being who, in order to control the process, would look at some aspect of the process and, if it weren't correct, would make an adjustment. For example if the temperature of a stream of water passing through a steam heat exchanger were required to be 50°C the operator would measure the temperature of the water and, if it were too high would reduce the steam flow to the heat exchanger by closing a valve or, if the temperature were too low, he would open the valve to increase the steam supply. Automatic systems work in the same way by measuring a characteristic of a stream – *the measured variable* – and comparing the value with what it should be – *the desired value* – and making an adjustment to a device – *the control element* – which adjusts some characteristic – *the controlled variable* – which affects the measured variable. The *control action* continues until the measured variable has the desired value. In the case above, the measured variable is the water temperature and the desired value 50°C. The control element is the steam control valve and the controlled variable the steam flow. The mathematics of the control loop are carried out by a specialist computer called a *programmable logic controller* or PLC.

AUTOMATIC CONTROL
Automatic control systems can ensure more consistent plant operation and maintain better product quality.

Section 7.4 requires a study time of about 30 minutes

7.4 Elements of Process Design

7.4.1 Approach to Design

The sort of water treatment plant with which process engineers are concerned vary from simple water filters costing a few thousand pounds up to complete water or sewage treatment works costing tens of millions of pounds and taking a year or two to construct. It is very important that mistakes do not occur and design is a team effort involving not only process engineers but also other disciplines including civil, mechanical, electrical and control engineers and may well include laboratory and pilot plant investigations to ensure that the selected process will work. The design process, then, has to be *auditable*, by which we mean that, at every stage, the basis of the design must be recorded together with any assumptions made. The design must be developed in a logical and systematic way and all calculations, no matter how apparently trivial, must be checked for errors.

DESIGN PROCEDURES
Communication is critical to engineering design.

Formalising the way in which calculations and drawings are prepared ensures that checks and audits can be carried out easily and that all members of the design team can be made aware of changes.

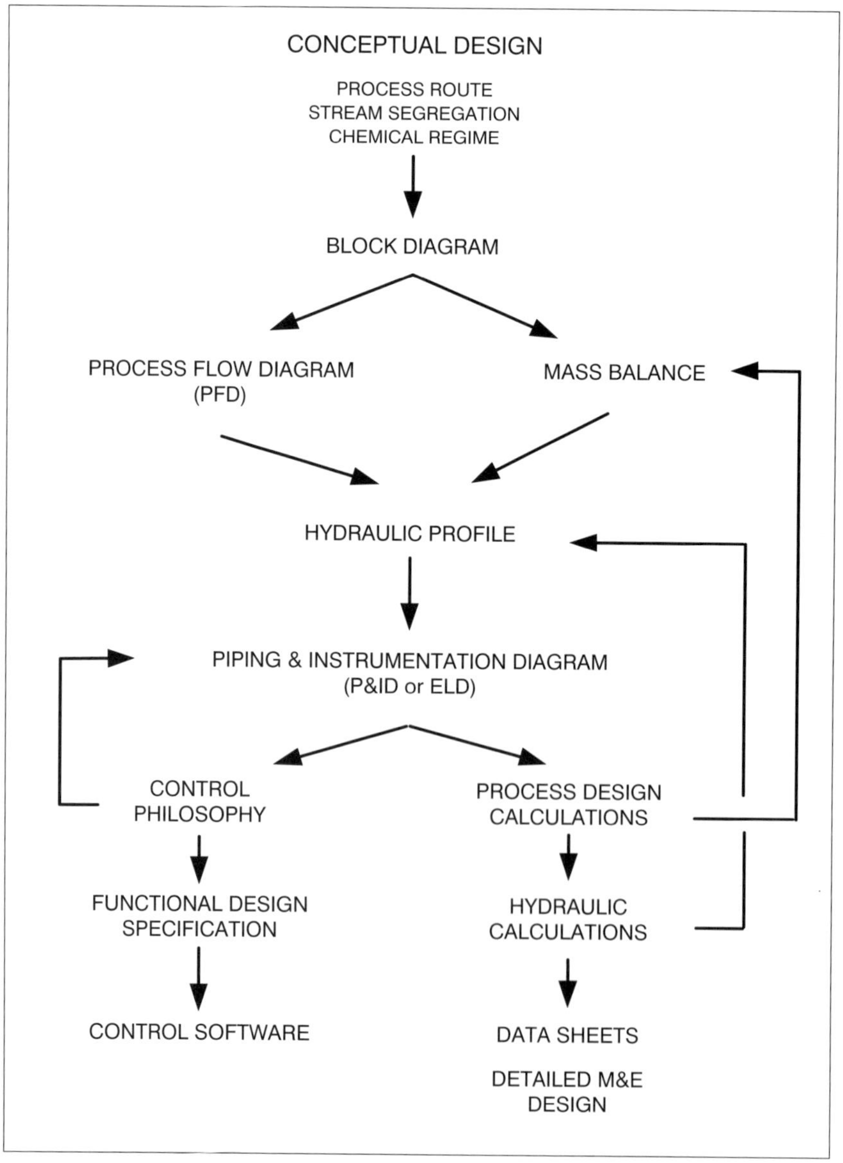

Fig. 7.2 – Process design documents

The process engineer's job is to ensure that the process – the sequence of unit operations – will satisfactorily convert the *raw material* – in our case the influent water – into the specified *product* or treated water quality. He or she will then be responsible for designing the individual unit operations and specifying dimensions and materials of construction so that a mechanical engineer or civil engineer can construct the plant. This will also include hydraulic design and specification of pumps and pipework. Finally it is the process engineer's responsibility to prepare a detailed *control philosophy* which explains exactly how the plant will be controlled in sufficient detail to allow a controls engineer to develop a *functional design specification* for the control system.

The sequence in which the design proceeds is shown in Figure 7.2 and the documents referred to are explained in the following sections.

7.4.2 Block Diagrams

BLOCK DIAGRAMS
Block Diagrams are a means of formalising and evaluating alternative conceptual designs.

The fundamental requirement before starting a design is to know what it is required to do. In our case this means that we need to know:

- the flow to be treated (average, maximum and minimum and diurnal and/or seasonal variations)
- the quality of the raw material (the raw water) which means not just an analysis of a sample and not just maximum, minimum and mean but information on how the quality varies diurnally and/or seasonally
- the required quality of the treated water – consent limits for a wastewater discharge, quality standards for potable or industrial water.

The first step in the design process is the preparation of a *flowsheet* or *block diagram*. This is a simple representation of the sequence of unit operations which will be needed in the plant and allows discussion and development. The block diagram in Figure 7.3 is for a simple water works treating reservoir water for potable supply. It will raise important questions like how to deal with wastewater from sludge thickening and dewatering operations and the quality and quantity of sludge for disposal. This may result in changes to the flowsheet – the first example of the continual review and updating which is characteristic of the design process.

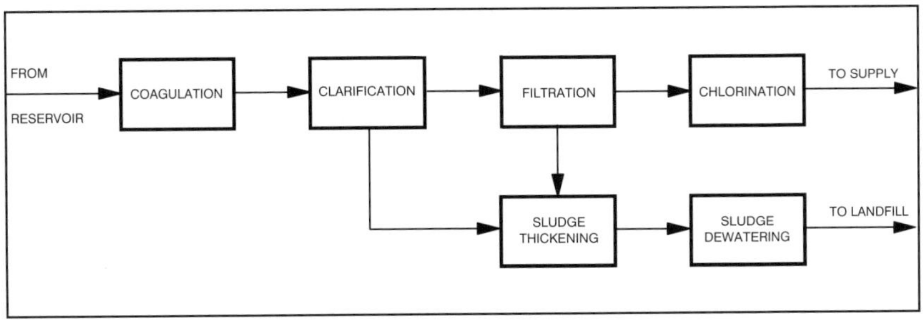

Fig. 7.3 – Block diagram

This is only one of a number of possible flowsheets and, at this stage, they will need to be evaluated in terms of:

- process performance (will it meet the required specification)
- capital cost (estimated $\pm 30\%$)
- operating cost (estimated $\pm 30\%$).

On the basis of this evaluation one flowsheet will be selected for design (estimated ±30%).

7.4.3 Process Flow Diagram

PFDs
Process Flow Diagrams identify the process route and the principal unit operations.

They may incorporate mass and heat balances.

Having decided on the flowsheet, the block diagram will be developed into a *Process Flow Diagram* or PFD. This will show, usually symbolically, the unit operations defining the type, and will also indicate where pumps will be required. Figure 7.4 shows a PFD developed from the block diagram in Figure 7.3. It now defines the type of clarifier (lamella settling) and the type of filter (rapid gravity sand filter) as well as the number of each.

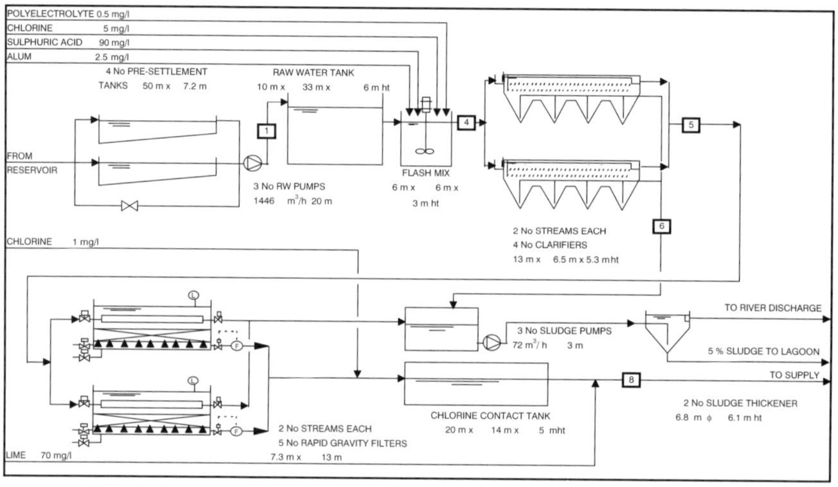

Fig. 7.4 – Process flow diagram

7.4.4 Mass and Heat Balances

MASS AND HEAT BALANCES
Mass and heat balances identify the physical, chemical and thermodynamic properties of each stream on the plant.

More information about the quality and quantity of each stream on the plant is now needed. This involves the preparation of mass and, possibly, heat balances so that the mass flow, composition (usually expressed as the concentration of the various components, and thermodynamic data can be identified for each stream on the plant. The calculations involved will be explained in Unit 8 of this Module and it is usually convenient to present the results on the PFD, either in a box adjacent to the stream line or, more usually, in tabular form below the diagram itself.

7.4.5 Hydraulic Design

The possible need for pumping has been mentioned, and it is important that this need is identified by conducting calculations to assess the head losses through the various parts of the plant including pipework. This is demonstrated in Unit 11 of this Module.

When those hydraulic calculations have been performed the results will be presented in the form of an *hydraulic profile* through the plant. Figure 7.5 shows the hydraulic profile for the plant shown in Figure 7.4.

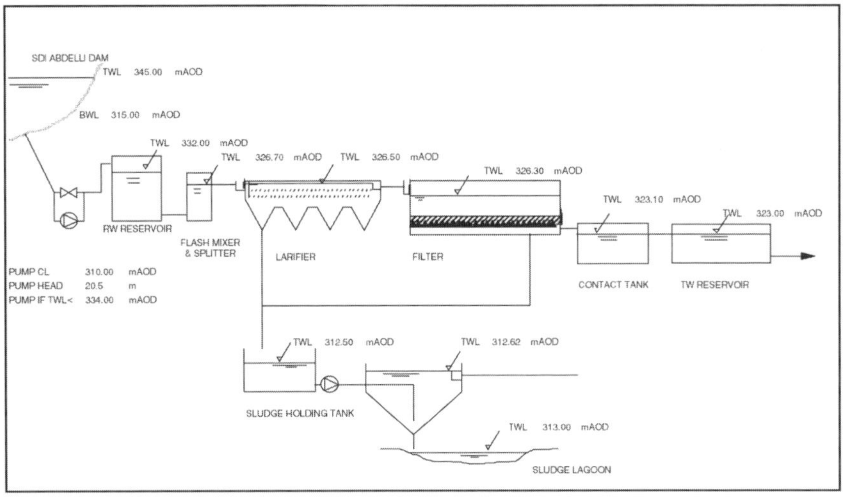

Fig. 7.5 – Hydraulic profile

HYDRAULIC PROFILE
The hydraulic profile identifies the need for pumping.

Clearly there is a good deal of interrelationship between the hydraulic profile and the PFD and it is not unusual for the hydraulic profile to identify a need for additional pumping which will mean the PFD has to be revised.

7.4.6 The P&ID

The Piping and Instrumentation Diagram (P&ID) or *Engineering Line Diagram (ELD)* is the key engineering drawing in any project because it completely defines the process plant. Until the P&ID is finalised or *frozen*, no detailed engineering design can be undertaken. In this sense it represents the passing of engineering responsibility from the process engineer to the mechanical, electrical and civil engineers – although, in reality, the process engineer continues to have an input into the plant until commissioning is completed.

P&IDs
The P&ID is the most important drawing produced during the design of a treatment plant because it is the basis for all mechanical, electrical and control system designs.

The P&ID is a schematic representation of the plant but in much more detail than that given in a PFD in that it will show every item of the plant: valves, pipes pumps, unit operations even temporary commissioning items like strainers inserted into pipes to protect pumps from construction "residues" – nuts, bolts, weld splatter and so on. Whereas a PFD might indicate a pump, the P&ID will identify duty and standby pumps together with all the isolating valves, non return valves and relief valves. The P&ID is not a scale drawing and is not intended to show how the items of plant are laid out, only how they are connected together and controlled. It will show:

- every item of equipment identified by a unique *tag number*
- every section of pipe identified by a line number, size and material
- every valve identifying type and size
- every control loop identified by a tag number
- each individual instrument in that loop identified by a tag number relating it to the loop.

In order for a P&ID to be interpreted it is necessary to use standard symbols. Unfortunately there is no single standard and, although there are British Standards available, they do not include all unit operations used in water treatment so that engineers, in Europe and the USA, have developed non-standard but widely used symbols. Figure 7.6 shows some typical symbols used in P&IDs and Figure 7.7 gives an example of their use.

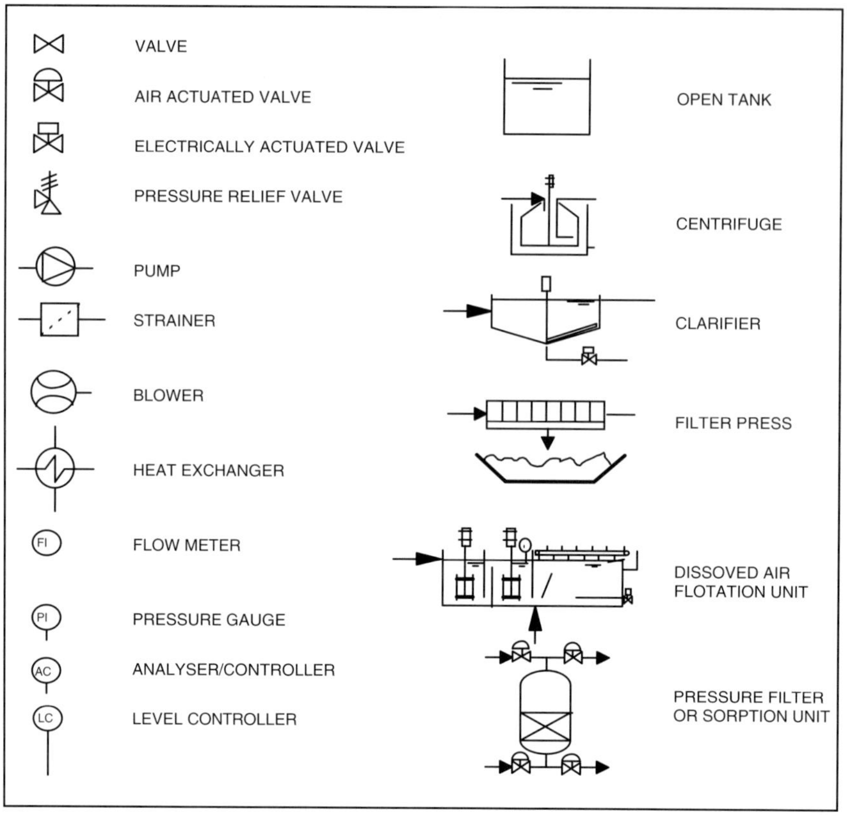

Fig. 7.6 – Typical graphical symbols

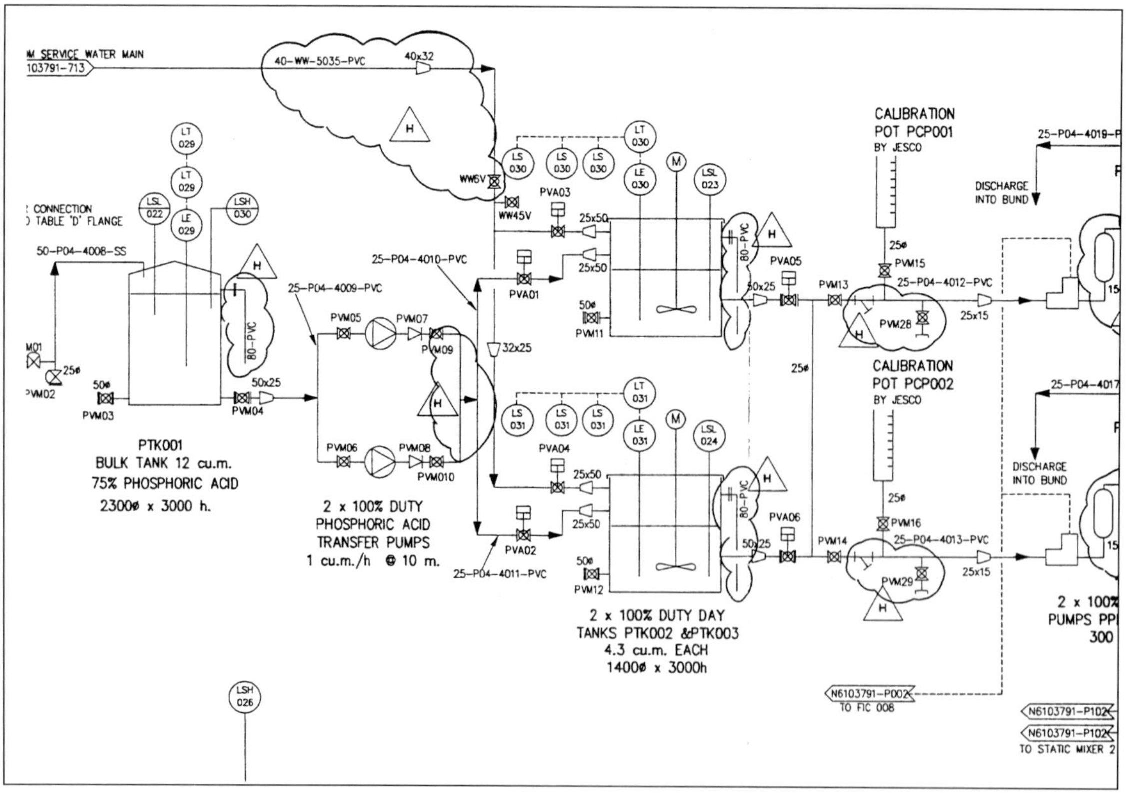

Fig. 7.7 – A section of a P&ID

7.4.7 Schedules and Data Sheets

Once the calculations P&ID has been drafted, other documents can be prepared. The most important are equipment, valve, pipe and instrument schedules. These are, essentially, shopping lists which identify each component on the P&ID by Tag Number and provide a one-line description and specify size, material, type and supplier's model number. This will be an important reference document when it becomes necessary to replace items in the future.

When the individual process units such as pumps, filters, reactors and so on have been designed the engineer will complete a Data Sheet (Figure 7.8). This is a brief specification which gives sufficient detail about the duty, size, materials and so on of the item of equipment to allow it to be described to a supplier so that it can be purchased.

EQUIPMENT DATA SHEET

Equipment name		Borehole 2 pump
Tag number		P02
Number		1
Type		Submersible multistage centrifugal
Supplier		Ingersoll Dresser Pumps
Supplier's model number		PN81
Delivery flow	m^3/h	42
Delivery head	m	80
Suction condition		Flooded
Suction connection	mm	N/A
Delivery connection	mm	150 Flanged PN16
Fluid pumped		borehole water
Solution strength	%	N/A
Specific gravity	SG	1.0
Viscosity	cP	100
pH		6-8
Suspended solids	mg/l	Nil
Temperature	°C	ambient
Material-head		cast iron
- impeller		aluminium bronze
- shaft		carbon steel
Seal type		mechanical
Drive type		submersible
Power rating	kW	15
Speed	rpm	2900
Power supply		415v 3ph 50Hz
Vendor to include		Anti-condensation heater

*information by vendor

					PROJECT :
2	23/04/97	HEAD REVISED			SOUTH STAFFORDSHIRE WATER
1	16/04/97	FOR ENQUIRY	RWH		SLADE HEATH MANGANESE
0	24/03/97	ORIGINAL ISSUE	RWH		REMOVAL PLANT
REV	DATE	DESCRIPTION	BY	CHKD	REF : REF : 8692-002.SP

Fig. 7.8 – Typical equipment data sheet

7.5 Self Assessment Questions

SAQ 7.1 The Fanning equation for pressure loss due to pipe friction is:

$$\frac{\Delta P}{l} = \frac{4fv^2}{2gd}$$

where P = pressure loss
l = length of the pipe
f = friction factor
v = velocity
g = acceleration due to gravity
d = pipe diameter

What are the dimensions of the friction factor, f?

SAQ 7.2 What is the absolute pressure in kPa at the bottom of a tank containing a depth of 8 mg of water?

SAQ 7.3 A sewage treatment works produces 120 m^3 per day of sludge which contains 2% by weight of solid material. If the sludge density is 1050 $kg.m^{-3}$ what is the mass of solid material produced in tonnes per day?

SAQ 7.4 A pump imported from USA is rated to deliver 600 gpm (gallons per minute) of water at a head of 45 ft. What is the power absorbed in kW?

Density of water = 1000 $kg.m^{-3}$
g = 9.81 $mg.s^{-2}$
1 W = 1 Joule/s

SAQ 7.5 A filter at a UK waterworks which was constructed in 1955 is 9' 6" wide × 17' 9" long and was designed for a hydraulic loading (flow divided by cross sectional area) of 1.5 gpm/ft^2. How much water can it filter in Mld (i.e. Megalitres per day)?

SAQ 7.6 The normal units quoted for specific energy demand are KWh per cubic metre. Assuming a density of 1000 kg/m^3, what is the equivalent in SI units of m and s? How does this relate to pressure per unit density?

7.6 Nomenclature

D diffusion coefficient, $m^2.s^{-2}$

d diameter, m
g gravitational constant, 9.81 $m.s^{-2}$
k mass transfer coefficient, $m^2.s^{-2}$
m mass, kg
v velocity, $m.s^{-1}$

μ viscosity, $kg.m^{-1}.s^{-1}$
ρ density, $kg.m^{-3}$
σ surface tension, N/m

7.7 Solutions to Exercises

Ex 7.1 v has dimensions LT^{-1}
Q has dimensions MT^{-1}
D has dimensions L
LHS has dimensions LT^{-1}
RHS has dimensions $MT^{-1} \times L^{-2} = MT^{-1}L^{-2}$
by inspection we need to multiply RHS by L^3M^{-1}
or divide by ML^{-3}, which is density, so equation should be:
$v = Q/\rho D^2$

Ex 7.2 Volume flow = 50 m³/day = 50/24 = 2.1 m³/h
Density = 1020 kg/m³
Mass flow = 2.1 × 1020 = 2140 kg/h

Ex 7.3 Mass = 2 te = 2000 kg
Force = 2000 × 9.81 = 19,620 N
Distance = 7.5 m
Work = 19620 × 7.5 = 147150 N.m = 147150 J

Ex 7.4 Mass flow = 50 m³/h × 1000 kg/m³ = 50000 kg/h
Mass flow = 50000 kg/h / 3600 s/h = 13.9 kg/s
Distance = 5 m
Power = 13.9 kg/s × 9.81 m/s² × 5 N.m/s = 682 J/s or W
So, Power = 682 W = 0.68 kW

Ex 7.5 $P_{abs} = 50 \times (1/2.2)\ kg/\ (2.54^2\ cm^2) = 3.52\ kg/cm^2$

Ex 7.6 75 m³/h = 75/3600 = 0.020833 m³/s

PIPE TYPE	ID, mm	AREA, m²	VELOCITY, m/s
4" ID	101.6	0.008107	2.569694
4" NB PVCU Class C	100.0	0.007854	2.652582
4" NB PVCU Class E	93.1	0.006808	3.060339
110 mm OD PVCU	93.6	0.006881	3.027730
110 mm OD Hep_3O – max	104.2	0.008528	2.443056
110 mm OD Hep_3O – min	102.8	0.008300	2.510052
Stanton DI Class K9	87.8	0.006055	3.440962

8. Mass and Heat Balances

Contents

Aims and Objectives

This unit is an introduction to Mass and Heat Balances, a thorough understanding of which is important for your understanding of future Modules in the series *Water and Wastewater Process Technologies*.

After studying these notes you should be able to:

1. explain in your own words the following terms and concepts:

   System
   Mass balance
   Heat balance

2. carry out mass and heat balance calculations including systems with:

   recycle
   accumulation
   chemical reaction

It is important that you are able to correctly complete all the self assessment questions at the end of these notes.

Essential Prerequisites

Prior to undertaking this lecture it is important that you have completed and understood the following lectures from the *Process Science and Engineering* Module:

Unit 1 Fundamentals of Water Chemistry
Unit 2 Chemical Kinetics and Equilibria
Unit 7 Fundamentals of Process Engineering

8.1 Introduction

MASS BALANCE
What goes in must come out unless it stays there.

Mass and heat balances are simple but very powerful design tools developed by process engineers from the laws of conservation of matter and heat to allow the quantitative description of a particular process or system. A *system* is a volume of space enclosed by an arbitrary boundary inside of which is a *process* causing a change in state or chemical composition of matter. This may be a simple blending of two streams of water at different temperatures to produce a flow of water at a third temperature, as in a shower, or it may be a chemical reactor in which chemicals react to form products. Mass and heat balances are the process engineer's equivalent of an accountant's balance sheet, predicated on the very simple principle that *what goes in comes out unless it stays there.*

Mass and heat balances are also referred to, respectively, as *material balances* and *enthalpy or energy balances*. In the water industry, it is usual to think in terms of volumetric concentration of materials; e.g. in meeting the potable water quality standard for aluminium, its concentration should be $<200\,\mu g/l$; a discharge consent to the environment may include a 20 mg/l BOD standard. These are useful numbers in defining input/output qualities but volumes may change in chemical processes so volumetric concentrations are not a reliable measurement for mass balance calculations and so mass flows in mass per unit time must be considered.

Section 8.2 requires a study time of about 30 minutes.

SYSTEMS
Open systems can exchange mass and/or energy with their surroundings. Closed systems cannot.

8.2 Processes and Systems

8.2.1 Systems

Mass and heat balances can be set up around a *system* (Figure 8.1) – the thermodynamics term used to identify any process to be investigated. The system will have a boundary with the surroundings. If material and energy cannot be exchanged with the surroundings, the system is *closed* and has constant mass and energy. If material and/or energy can be exchanged with the surroundings, it is *open*. A system can be as large and complex as a complete factory, or as small as a pipework tee.

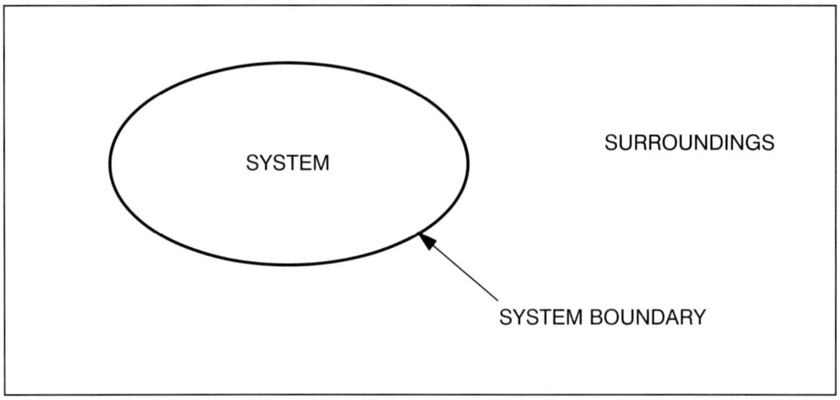

Fig. 8.1 – A thermodynamic system

When considering the system there is no need to be concerned, at least for the time being, with how changes occur *within* a system but only their effects on the surroundings and in particular, their effect on other systems joined to the one being considered. Therefore mass balance is extremely useful at examining or defining the inputs and outputs of a particular process, or unit operation, and potential impacts on other unit operations within a flowsheet. It can also be used to assess total inputs and outputs from a whole flowsheet, or any part of it around which a boundary is drawn.

8.2.2 Processes

PROCESSES
The three common process types:

- Batch
- Fed batch
- Continuous.

Processes cause changes in the system or surroundings. There are three common types of processes in water and wastewater treatment:

Batch Processes: All materials are added at the start, the system is closed, then products are removed at the end of the reaction.

Fed-batch: Materials are added to material already in the reactor, the system is closed, then products (that is the whole reactor contents) are removed at end of the reaction.

Continuous: Material flows in and out of the system continuously.

Batch process tend to be used when volumes requiring treatment are small or the product has high added value. Many sludge treatment processes in the water industry are batch, e.g. filter presses. Other sludge process are effectively continuous but are often mistakenly called fed-batch – they are in effect semi-continuous, as products are removed at intervals, the same time as material is added. In water and wastewater treatment more often than not we are dealing with continuous systems at *steady state*, e.g. sedimentation, percolating filters. Continuous processes are usually operated at steady state.

8.2.3 Steady State

A process is said to be at steady state if none of its properties, e.g. concentration, temperature, pressure, vary with time. Continuous processes, the most important in water and wastewater treatment flowsheets, are usually operated at, or as close as possible to, steady state. Batch and fed-batch processes are transient and therefore operated under *unsteady state* conditions, i.e. changes in properties of the system do occur with time, e.g. mass of an input reduces with time. Continuous processes can also be operated under unsteady state conditions.

STEADY STATE
A system is at steady state if none of its properties changes with time.

Steady state is not the same as *equilibrium*. If a system is at equilibrium, there is no overall change in that system and its surroundings with time. This clearly is not what is required of a process or unit operation, therefore a process is operated to avoid equilibrium by providing a driving force. This can be achieved by continuously disturbing the system, e.g. heating or adding raw material. This happens in continuous processes operated at steady state – mass is being added to and taken out of the system all the time, so that an overall change in the system and its surroundings can take place.

STATE OF PROCESS
Equilibrium = No net rate
Steady state = no change in rate
Unsteady state = change in rate

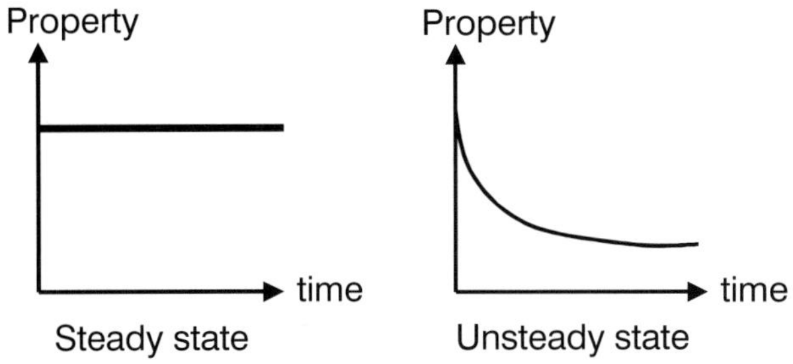

Fig. 8.2 – Comparison of steady and unsteady state processes

Section 8.3 requires a study time of about 2 hours

STEADY STATE BALANCE
For a continuous process at steady state:
mass flow in = mass flow out
but
volume in ≠ volume out.

Ex 8.1
Which of the following units are suitable for a mass balance?
1. *m^3/h*
2. *kg/s*
3. *mg/l*
4. *kmol/day*
5. *l/s*

Select a system boundary which allows simple calculation.

FLOW DIRECTION
Always include an arrow to indicate direction of flow into and out of system boundary.

MATERIAL
Select the material for the balance carefully.

UNITS
Ensure that the units of measurement used in mass balances are consistent.

8.3 Mass Balance Calculations

8.3.1 The Basic Principles

Mass balances are based on the simple criterion that mass is conserved so, for a continuous process at steady state:

$$\textbf{mass flow in} = \textbf{mass flow out} \qquad \textbf{(8.1)}$$

In most process engineering problems the starting point will be a flowsheet of some sort. Consider the very simplest of water treatment processes: a solids separation clarifier. Here water containing suspended solids passes through a settling tank where some of the solids are removed as a *sludge* that is then disposed of. To estimate the sludge disposal costs the amount of it being produced must be calculated. The flow sheet (Figure 8.3) is very simple:

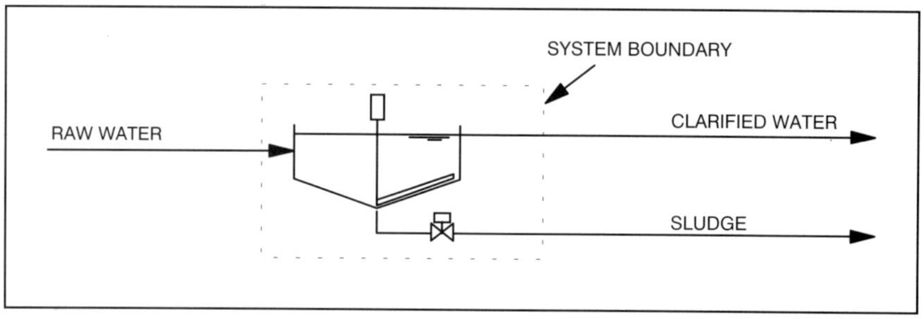

Fig. 8.3 – System boundary

The dotted line represents the boundary within which the mass balance is to be performed. As already stated, the system boundary can be drawn around a single unit of equipment such as a tank or pipe-work tee, several unit processes or the whole plant. The most important thing is that the boundary is a closed envelope, that is it completely encircles the system. There are three streams crossing the boundary in Figure 8.3: a single stream entering the system and two streams leaving it. There is no need to consider in detail what happens inside the boundary except for how it affects the three streams.

A mass balance is influenced by the properties of the material under consideration. Most water treatment process mass balances will be based on water and the contaminants in it. However, because water contains so many contaminants, it is not practicable to consider all of them. In this case it is only the suspended solids, as well as the water, that need to be considered. In other cases other materials may need to be included on the balance. A mass balance can be written for the total materials involved; any unchanged species (for example nitrogen) or an individual element, whether reacted or not, e.g. carbon.

8.3.2 Defining the Flows

It is essential that mass balances calculations are carried out in consistent units – all kg/h, lb/h, tonne/day or whatever is chosen. If the system is at steady state, the simple rates of mass flow can be used. This is a *differential balance*. However, in many water treatment processes, some flows will be intermittent so the *instantaneous* flow may be high but the *average* flow small. For example clarifiers are used to separate suspended solids from water to form a sludge.

Sludge is usually removed from a clarifier by intermittently opening a valve which allows sludge to discharge. Suppose the sludge is discharged at a flow of 25 te/h for 2 minutes every 30 minutes. The total discharge in any hour is ($2 \times 25 \times 2/60$) = 1.67 te, that is an *average* flow of 1.67 te/h.

A mass balance based on the total mass over a fixed period of time is called an *integral balance*. In water and wastewater treatment, both differential and integral approaches can be used but as most processes usually operate under non-steady flows, an averaged rate or total mass over time needs to be used. It is normal to use this average flow in mass balances unless the rate of accumulation (in this case of sludge) is of interest.

INTERMITTENT FLOW
Convert intermittent flows into pseudo continuous flows by averaging the total flow over the total time period.

Ex 8.2
Water is pumped to a tank by a pump which operates twice each hour for ten minutes at a flow rate of 20 l/s. What is the average hourly flow into the tank in m^3/h?

8.3.3 Balance Calculations

In this example, if the mass flow of raw water into the system is R kg/day, the mass flow of water in the waste sludge is S kg/day and the mass flow of clarified water is Ckg/day, then:

$$R = S + C \qquad (8.2)$$

The mass flow of any contaminant in water is the flow of water multiplied by the concentration of the contaminant. So, if the raw water suspended solids concentration is r mg/kg, the clarified water suspended solids is c mg/kg and the suspended solids in the sludge is s mg/kg, the mass balance for the suspended solids in mg/day can be written as follows:

$$R \times r = (S \times s) + (C \times c) \qquad (8.3)$$

MASS FLOW
The mass flow of any species in water is given by
mass flow = flow × concentration

WEIGHT PERCENT
Weight percent
= weight solids/(weight solids+liquid)
A 5 wt % solids expressed as a concentration is 0.05 kg/kg or 50,000 mg/kg.

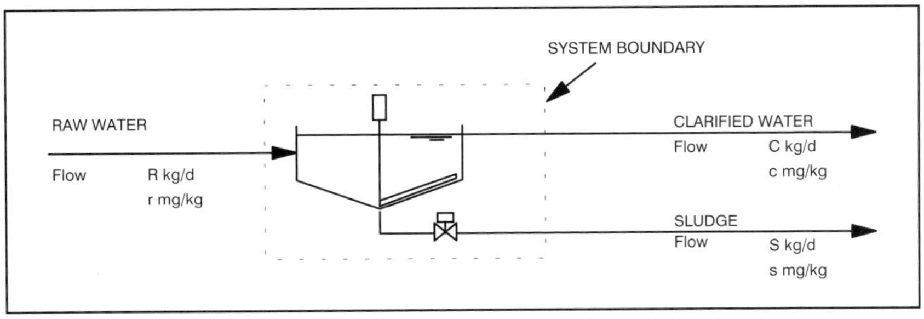

Fig. 8.4 – Mass balance calculation

Ex 8.3
A wet sludge is found to contain 71% of water by weight. After drying, 60% of the original water has been removed.
Calculate the weight of water removed per kg of wet sludge and the solids content (% by weight) of the dried sludge.

Suppose that a water flow of 2 Mld (megalitres/day) of very muddy river water containing 300 mg/l of suspended solids is to be treated. Let it additionally be assumed that the clarifier leaves 20 mg/l of suspended solids in the clarified water and that the sludge product has a suspended solids concentration of 0.5% by weight. How much sludge is produced and how much clarified water?

Concentrations in water treatment are traditionally measured in mg/l which, as stated earlier, are not appropriate units for mass balances. Fortunately water has a density of 1000 kg/m^3 so it is easy to convert flows in, say, m^3/h to kg/h and concentrations from mg/l into mg/kg.

Ex 8.4
Sea water of density 1.01 g/ml containing 28 g/l of sodium chloride is evaporated to produce salt. If the salt production required is 2 tonne per day and the plant operates for 8 hours each day, what is the sea water flow (mass and volume) which has to be processed?

The inlet raw water flow, R, is 2 Mld, that is 2,000,000 kg/day with a suspended solids concentration, r, of 300 mg/l – or 300 mg/kg. S kg/day of sludge is produced having concentration 0.5%, that is 0.5 g/100 g or 5 g/kg which is the same as 5000 mg/kg. The clarified water flow will be C kg/day with a suspended solids concentration of s mg/kg. Equations 8.2 and 8.3 give us:

$$2{,}000{,}000 = S + C$$

$$2{,}000{,}000 \times 300 = S \times 5000 + C \times 20$$

which we can solve simultaneously to give C = 1,887,550 kg/day (1.88 Mld) and S = 112,450 kg/day of sludge for disposal (Figure 8.5).

Ex 8.5
10 m^3/h of water containing 80 mg/kg of nitrate is mixed with surface water containing 10 mg/kg of nitrate. If the mixed stream is to have a nitrate concentration of no more than 50 mg/kg what is the minimum flow of surface water required?

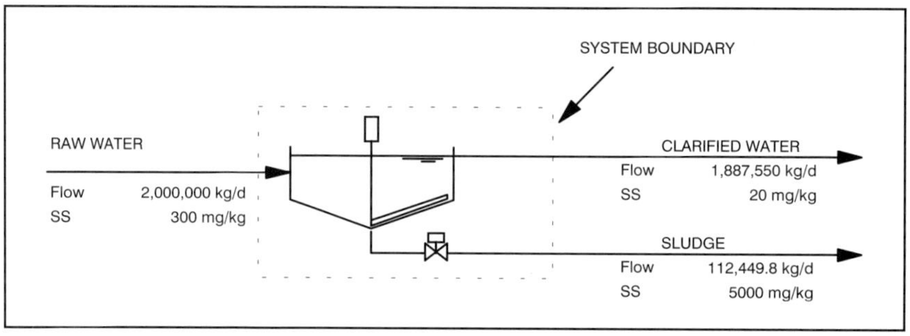

Fig. 8.5 – Mass balance

8.3.4 More Complex Systems

COMPLEX SYSTEMS
Split complex flowsheets into sub-systems which can be more easily managed.

More complex flow sheets can be broken down in sections. For example, suppose that the sludge in the example above is dewatered to 40% solids by a plate and frame filter press with press filtrate disposed of to waste. The flow sheet is shown in Figure 8.6.

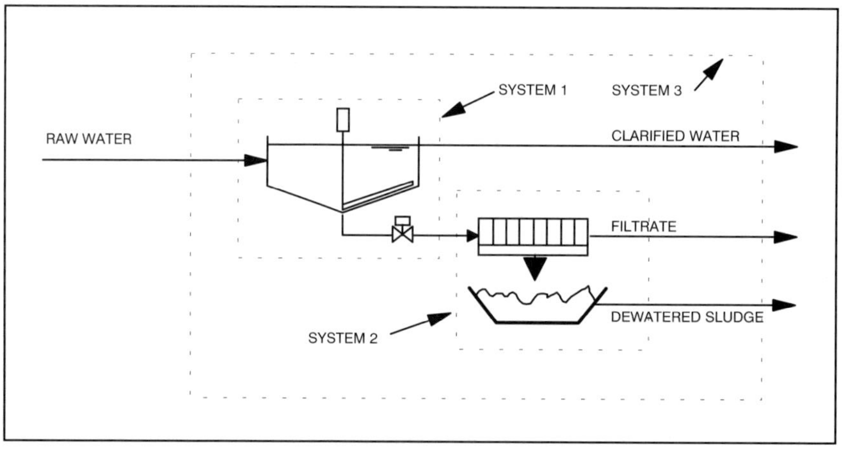

Fig. 8.6 – Clarification with sludge dewatering

Ex 8.6
The sludge from the process shown in Fig 3.3 is to be dewatered to 40% dry solids by filter pressing. Calculate the mass (in kg) of 40% dry solids sludge produced per day and the volume of filtrate (Figure 8.6) assuming that it contains 100 mg/l of suspended solids.

The mass balances around the clarifier (Figure 8.6 System 1) can be carried out separately from the filter press (System 2). A mass balance carried out around the complete plant (System 3) should yield the same result as the two individual mass balances.

For very complex flowsheets it will normally be necessary to carry out a series of mass balances around various sections of the plant and the choice of a suitable boundary for the mass balance will be made on the basis of the information that is available or can be predicted for the various streams.

8.3.5 Recycle Streams

Mass balances become quite difficult when there is a recycle stream involved. Suppose that, in the example above, the filtrate from the filter press, containing, say, 100 mg/l of suspended solids, were to be recycled to the clarifier inlet instead of going to waste (Figure 8.7). An overall mass balance around System 3 can still be conducted but this then affects Systems 1 and 2 since the mass flows of water and suspended solids into the clarifier have increased because of the recycle stream. This means that the clarified water and sludge flows will change, which means that the sludge load to the filter will change, which will change the recycle stream. A loop is created which makes the calculation much more difficult.

RECYCLE FLOWS
Mass and heat balances involving recycle streams generally require iterative calculations.

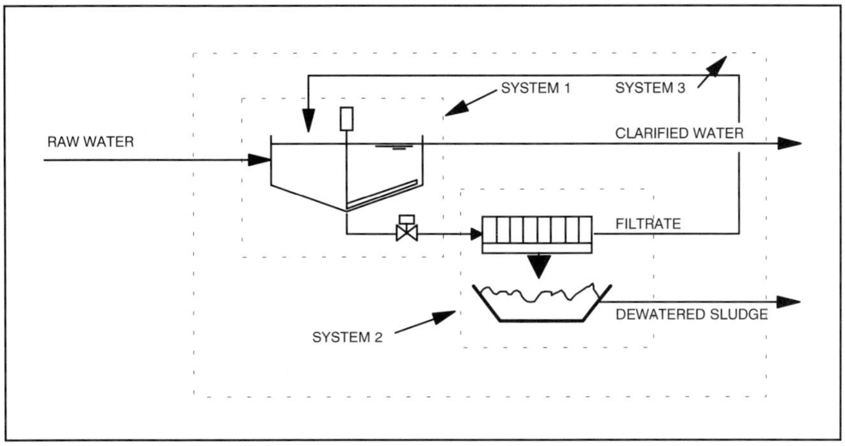

Fig. 8.7 – Flowsheet with recycle

In this situation it is often necessary to make a guess at the recycle flow and then carry out a number of iterative calculations around the loop until *convergence* occurs, i.e. until the calculated recycle flow is sufficiently close to the previously calculated recycle flow. In the case of the filter press filtrate recycling, the contribution of the recycle flow is so small that it could be ignored, but that is not always the situation. The easiest solution to recycle problems is to set up a spreadsheet solution and let the computer carry out the iterative calculation.

8.3.6 Accumulation

If the system is not at steady state there will often be a difference between system inputs and outputs. This means that there is either a net accumulation of material (for example a balance tank filling) or there is a net loss (for sludge being removed). The rule here is:

ACCUMULATION
For a system in which there is accumulation or loss of material
mass in − mass out = accumulation

What goes in minus what comes out must still be in the system

mass flow in − mass flow out = rate of accumulation (8.4)

Ex 8.7
Water is pumped from the tank in Ex 1 at a constant rate of 23.5 m^3/h. What is the rate of accumulation? If the tank has a working volume of 12 m^3, how long will it be before the tank overflows?

STEADY STATE
For a system in which there is chemical reaction
mass in + mass generated = mass out + mass consumed

Ex 8.8
Hydrochloric acid reacts with lime as follows:

$$2HCl + Ca(OH)_2 = CaCl_2 + 2H_2O$$

How much lime in kg/h is required to neutralise 100 m^3/h of dilute hydrochloric acid containing 5% HCl if the acid solution has SG 1.023?

At Wts:		
	H	*1.0*
	Cl	*35.5*
	Ca	*40*
	O	*16*

Note remember to convert to mass rates in units of mols per unit time before applying reaction conditions and then convert back to initial units.

Ex 8.9
What will be the concentration of the calcium chloride solution produced in Ex 8.7?

8.3.7 Chemical Reaction

The generalised mass balance statements above apply when there is no kind of chemical reaction occurring within the system, i.e. no transformation of mass. Should there be one or more chemical reactions involved, this means that mass of a particular component may be destroyed or created within the system. Therefore a generalised mass balance with reaction can be written as:

"Input minus output plus generated plus consumed = accumulation"

Accumulation can be negative as well as positive. If the system is at steady state, then there will be no accumulation in that system so the general steady state mass balance equation can be written as:

mass in + mass generated = mass out + mass consumed (8.5)

A simple example is precipitation. A borehole water containing 10 mg/l of dissolved iron (Fe^{++}) is aerated to precipitate ferric hydroxide and then clarified to reduce the residual suspended solids to 5 mg/l (Figure 8.8).

The chemical reaction is (see also Unit C1, Section 6.1):

$$Fe^{2+} + 3H_2O + 1½O_2 = 2Fe(OH)_3$$

From the above reaction stoichiometry it can be calculated that the 10 mg/kg of Fe^{2+} will produce 19 mg/kg of suspended solids, most of which will be removed in the clarifier (Figure 8.8).

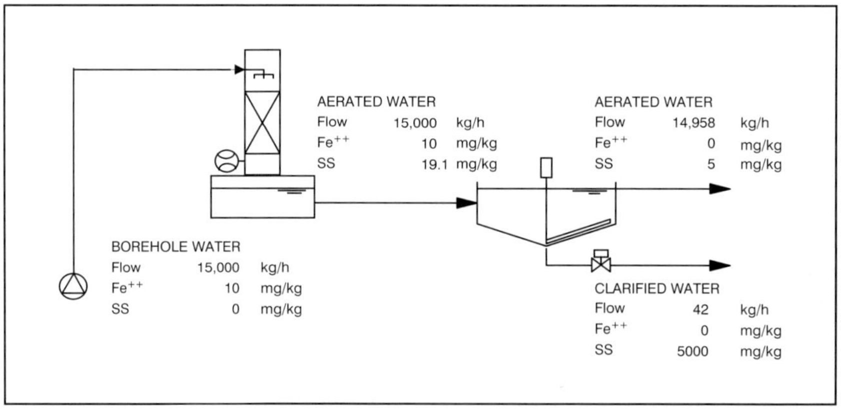

Fig. 8.8 – Chemical reaction

8.4 Heat Balance Calculations

Section 8.4 requires a study time of about 1 hour

8.4.1 The Basic Principles

Heat balances are based on the same premises as mass balances, namely that what goes in comes out. In this case it is not the contaminant concentrations that are of concern but the heat content or enthalpy. The heat balance is thus conducted in units of J/hr, rather than kg/hr.

Heat balances are not widely used in water treatment, although they are important in distillation, evaporation, sludge digestion and in some membrane applications. Generally heat balances are simpler than mass balances because there is only one stream characteristic (enthalpy) which is usually easy to define.

8.4.2 An Example

Ex 8.10
A flow of 5 m^3/h of hot water at 50°C is mixed with 10 m^3/h of cold water at 5°C, what is the temperature of the mixed stream?

Consider the simple blending of two streams of water, one of 10,000 kg/h at 10°C and the other 20,000 kg/h at 50°C. A mass balance tells us that the combined flow will be 30,000 kg/h at a temperature of T°C. In order to calculate T the heat content, or enthalpy, of each stream is required.

Enthalpy is measured in kJ/kg and, for water below its boiling point, is calculated as the temperature in °C multiplied by the mechanical equivalent of heat (4.2 kJ/kg.°C). The enthalpy of steam can be found from standard tables.

Ex 8.11
Calculate the power demand for an in-line electric heater to heat 1000 kg/h of water at 10°C to 40°C. Assume an overall efficiency of 90% for the heater element.

So the heat balance looks like this:

$$(10{,}000 \times 4.2 \times 10) + (20{,}000 \times 4.2 \times 50) = 30{,}000 \times 4.2 \times T$$

from which it can be calculated that T = 36.7°C.

8.4.3 Organising the Calculations

The key to setting up a heat or mass balance is to **be organised**. Although the following steps may seem over complex for trivial problems, it is good practice to apply a systematic approach which can be audited. The procedure is outlined in more detail in Doran (1995), Chapter 4.

METHODOLOGY
Always adopt a systematic and auditable approach to calculations.

There are four steps to mass or heat balance calculations:

- Assembly
- Analysis
- Calculation
- Finalisation

Each step is considered in more detail below:

8.4.4 Assembly

Draw a flow diagram – a simple box diagram is all that is needed – with quantitative information given on the diagram, *including units*. The quantitative information of interest is:

- masses
- mass flow rates
- mass compositions

The boundary of the system to be considered should be drawn on the flow diagram. Note that this boundary does not have to encompass a single unit operation, but can encompass as many or as few as is practical to achieve the required analysis.

Units used must always be consistent. Confusion often arises through mixing SI units with non SI units.

8.4.5 Analysis

Your analysis should state clearly:

Assumptions	You may not have all the information required, so you will need to use "engineering judgement" to make certain assumptions.
Additional data	Identify any additional data required to complete the calculations – if it is not available make suitable assumptions.
Basis	Clearly state the basis on which the balance is based: for continuous processes at steady state, select a specific time period and calculate the quantities of materials entering and leaving in that time period; for a batch or semi-batch process, the total quantities of materials added and withdrawn from the system should be used.
Chemical reactions	Make a list of reactants and products involved in reactions – there may be none but, if there are

any reactions you will need to know the formulae and molar masses of all the chemicals involved.

Equations

Identify and write down the appropriate mass balance equation in the form of equation (8.1), (8.4) or (8.5) in Section 8.3 above.

8.4.6 Calculation

Set up a calculation table showing all the components that pass in and out of the system or process, once again clearly show all units of measurement so as to avoid mistakes. Such tables are normally divided into two parts: in and out. Rows indicate the particular influent and effluent streams; columns will show the components of each stream. All known quantities are entered into the table. The mass balance equation is then applied and the unknown quantities entered in the table.

8.4.7 Finalisation

Once answered, again make absolutely sure that the numbers and units are stated clearly with significant figures. Finally, check your results!

8.4.8 Presentation

Mass and energy balances are normally presented in the form of process flow diagrams (PFDs). Either each stream is annotated with a box giving flow and composition data, or the stream is identified by a stream number and stream numbers, flows and compositions are tabulated below the flow diagram.

8.5 Self Assessment Questions

SAQ 8.1 Water is pumped to a tank by a pump which operates four times each hour for five minutes at a flow rate of 50 l/s. What is the average hourly flow into the tank in m^3/h?

SAQ 8.2 A wet sludge is found to contain 75% of water by weight. After drying, 50% of the original water has been removed.

Calculate the weight of water removed per kg of wet sludge and the solids content (% by weight) of the dried sludge.

SAQ 8.3 A solution of potassium chloride of density 1.12 g/ml containing 45 g/l of the salt is evaporated to produce a dry product. If the potassium chloride production required is 5 tonne per day and the plant operates for 12 hours each day, what is the solution flow (mass and volume) which has to be processed?

SAQ 8.4 100 m^3/h of surface water containing 110 mg/kg of sodium is mixed with borehole water containing 10 mg/kg of sodium. If the mixed stream is to have a sodium concentration of no more than 80 mg/kg what is the minimum flow of surface water required?

SAQ 8.5 50,000 kg/day of sludge containing 10,000 mg/kg of solids is to be dewatered to 40% dry solids by filter pressing. Calculate the mass (in kg) of 40% dry solids sludge produced per day and the volume of filtrate assuming that it contains 50 mg/l of suspended solids.

SAQ 8.6 Water is pumped at a constant rate of 15 m^3/h from a 30 m^3 capacity tank which is filled at an average rate of 20 m^3/h. What is the rate of accumulation? If the tank is empty, how long will it be before it reaches the overflow level?

SAQ 8.7 Sulphuric acid reacts with lime as follows:

$$H_2SO_4 + Ca(OH)_2 = CaSO_4 + 2H_2O$$

How much lime in kg/h is required to neutralise 350 m^3/h of dilute sulphuric acid containing 15% H_2SO_4 if the acid solution has SG 1.1?

At Wts:		
	H	1.0
	S	32
	Ca	40
	O	16

SAQ 8.8 The solubility of calcium sulphate is 2000 mg/kg. What will be the concentration of the suspended solids in the calcium sulphate stream produced in Q3.7?

SAQ 8.9 A laundry uses 300 m^3/day of good quality mains water for washing industrial workwear. Detergents and dirt from the wash loads add 600 kg per day of COD to the water. If 10% of the water is lost by evaporation, calculate the COD of the wastewater discharge in mg/l.

SAQ 8.10 New consent limits imposed on the laundry in Q3.9 will limit the COD concentration in the wastewater discharge to 1500 mg/l. It is proposed to treat the wastewater from the laundry by ultrafiltration, which is a membrane filtration process. Ultrafiltration splits the flow into two streams: 80% of the flow becomes filtrate and 20% reject. The COD concentration in the filtrate is 70% of that of the influent to the ultrafiltration unit and the removed COD is concentrated in the reject. Calculate the flows and COD concentrations of the filtrate and reject streams.

SAQ 8.11 The reject stream in Q3.10 will be treated by adding 270 mg/l of ferric chloride ($FeCl_3$) which will precipitate ferric hydroxide and remove 70% of the COD as suspended solids (2 mg/l COD = 1 mg/l suspended solids). The coagulated stream will be clarified in a settling tank which produces sludge at 2% solids and a clarified water stream containing 20 mg/l suspended solids. Calculate the clarified water and sludge quantities in kg/day.

SAQ 8.12 A reactor has a fresh inlet of 100 m^3/h with a COD concentration of 100 mg/l. The reactor achieves a 70% conversion of COD. What fraction of the flow must be recycled in order to ensure the effluent concentration is 10 mg/l. (This is an iterative solution best handled on a spreadsheet.)

SAQ 8.13 The laundry in Q3.9 intends to minimise wastewater discharge by dewatering the clarifier sludge to 35% dry solids and by recycling as much of the UF permeate as is possible for re-use within the laundry. Given the conditions set out in Q3.10 and Q3.11 calculate the maximum quantity of filtrate which can be recycled without exceeding the COD consent limit. (This is an iterative solution best handled on a spreadsheet.)

SAQ 8.14 Steel sheets are "pickled" by immersion in phosphoric acid after manufacture to remove mill scale and rust. The acid strength is reduced in the process and, when it falls to a set value, it is discarded. A laboratory test shows that 1 g of sodium hydroxide will neutralise 20 g of a waste pickling acid. Calculate the mass of slaked lime (kg), containing 94% of $Ca(OH)_2$, that is required to neutralise 1000 kg (1 te) of this liquor.

SAQ 8.15 A flow of 2 m^3/h of water at 8°C is heated to 60°C by direct injection of steam at 2bara in a special in-line steam blender. 3% of the steam is lost but the rest condenses. What is the required steam flow?

The enthalpy of steam at 2bara is 2707 kJ/kg.

SAQ 8.16 Calculate the power demand for an in-line electric heater to heat 1500 kg/h of water at 5°C to 100°C. Assume an overall efficiency of 90% for the heater element.

SAQ 8.17 A flow of 2 m^3/h of water at 8°C is heated to 60°C by direct injection of steam at 2 bara in a special in-line steam blender. 3% of the steam is lost but the rest condenses. What is the required steam flow?

The enthalpy of steam at 2 bara is 2707 kJ/kg.

SAQ 8.18 An evaporative cooling tower cools a circulating flow of 3000 m^3/h of water from 40°C to 35°C by evaporating a portion of the circulating water. How much water has to be evaporated to effect the cooling duty?

The enthalpy of steam at 35°C and atmospheric pressure is 2450 kJ/kg.

SAQ 8.19 In addition to the losses from evaporation a further 1% of the circulating water is lost through windage and blowdown. The losses are made up with mains water of chloride concentration 50 mg/l as Cl. What is the chloride concentration in the circulating water?

8.6 Nomenclature

- C mass flow kg/h, kg/day
- c mass concentration, mg/kg
- h specific enthalpy, kJ/kg
- H total enthalpy, kJ
- T temperature, °C

8.7 Solutions to Exercises

Ex 8.1 Ans 2,4 mass balances must be conducted in units of mass.

Ex 8.2 20 l/s = $20 \times 3600/1000 = 72\,m^3/h$
10 min = 10/60 = 0.167 h
Delivery in 10 min = $0.167 \times 72 = 12\,m^3$
Total delivery in 1 h = $2 \times 12 = 24\,m^3$
Average flow into tank = **24 m^3/h.**

Ex 8.3 1 kg sludge contains $1000 \times 0.71 = 710$ g water and $(1000 - 710) = 290$ g solids.
60% of the original water is removed, that is $0.6 \times 710 = 426$ g per kg of sludge.
Remaining sludge consists of $(710 - 426) = 284$ g water and the original 290 g solids. Total mass = 284 + 290 = 574 g so the solids content is 290/574 = 0.505 = **50.5%.**

Ex 8.4 Sea water density = 1.01 g/cm^3 = 1010 kg/m^3
Mass concentration = 28/1010 = 0.0277 kg/kg, or 27.7 kg/te
Put mass flow of sea water = Q te/h
Mass flow of salt = 27.7 kg/te $\times$ Q te/h
Mass flow of salt = 2 te/8 h = 2000/8 = 250 kg/h
$\therefore 27.7 \times Q = 250$
$\therefore$ Q = 9.02 te/h = $12.6 \times 1000/1010$ = **8.93 m^3/h.**

Ex 8.5

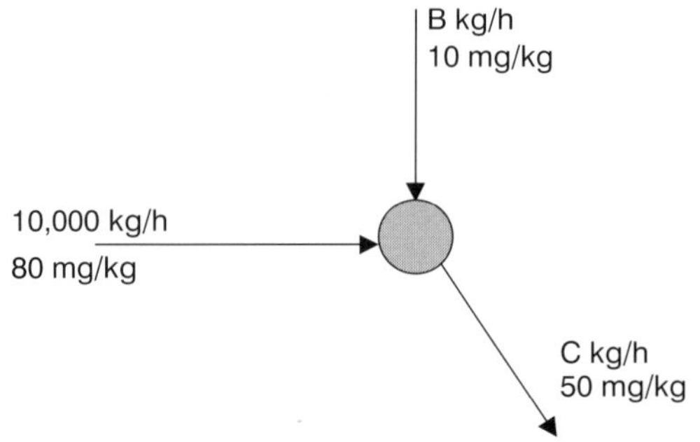

Write mass balance across mixing point as total flow and nitrate:

Equation 1: 10,000 + B = C
Equation 2: $10{,}000 \times 80 + B \times 10 = C \times 50$

Substitute Equation 1 into Equation 2:

$10{,}000 \times 80 + B \times 10 = (10{,}000 + B) \times 50$
$800{,}000 + 10B = 500{,}000 + 50B$
$40B = 300{,}000$
B = 7500 kg/h

Convert to volumetric flow rate = 7500/1000 = **7.5 m^3/h.**

Ex 8.6

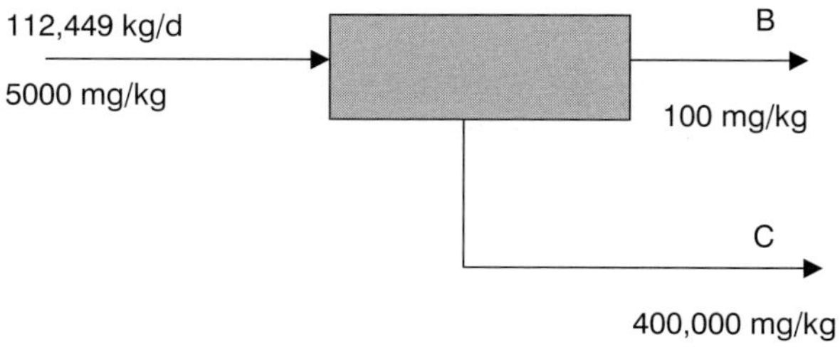

Write mass balance equations for total flow and sludge
Equation 1: $112{,}449 = B + C$
Equation 2: $112{,}449 \times 5000 = B \times 100 + C \times 400{,}000$

Eqn1 rearranged: $B = 112{,}449 - C$

Substitute into Equation 2:
$112{,}449 \times 5000 = 100(112{,}449 - C) + C \times 400{,}000$
$562{,}245{,}000 = 12{,}244{,}900 + 399{,}900\,C$
$C = 1378$ kg/d

Return to Equation 1:

$112{,}449 = B + 1378$
$B = 111{,}071$ kg/d

Convert to volumetric flow rate by density:
$B = 111{,}071/1000 =$ **111 m³/d**.

Ex 8.7 The average rate of flow into the tank is 24 m^3/h (Ex 1)
The average rate of flow from the tank is 23.5 m^3/h
Rate of accumulation = flow in − flow out
= 24 − 23.5 = 0.5 m^3/h
If the volume is 12 m^3 it will fill up in 12/0.5 = 24 hours.

Ex 8.8 Mass flow of acid solution = 100 × 1.023 = 102.3 te/h
Mass flow of HCl = 102.3 × 5/100 = 5.12 te/h = 5120 kg/h
2 kmol HCl require 1 kmol lime for neutralisation
2 × (1 + 35.5) = 73 kg HCl require 40 + 2 × (16 + 1) = 74 kg $Ca(OH)_2$
∴ 5120 kg/h HCl require 5120 × 74/73 = 5190 kg/h $Ca(OH)_2$
= 5190/1000 = 5.19 te/h.

Ex 8.9 In the reaction 73 kg HCl + 74 kg $Ca(OH)_2$ produce
111 kg $CaCl_2$ + 36 kg H_2O
So 5120 kg HCl + 5190 kg $Ca(OH)_2$ produce
5120 × 111/73 = 7785 kg $CaCl_2$ + 5120 × 36/73 = 2525 kg H_2O

Mass flow of acid solution = 102.3 te/h
Mass flow of HCl = 5.12 te/h
∴ Mass flow of water = 102.3 − 5.1 = 97.2 te/h
Mass flow of $Ca(OH)_2$ = 5.2 te/h
Total mass flow in = 97.2 + 5.1 + 5.2 = 107.5 te/h
Mass flow of water produced = 2525 kg/h = 2.5 te/h
∴ Mass flow of water leaving = 97.18 + 2.53 = 99.71 te/h
Mass flow of $CaCl_2$ leaving = 7785 kg/h = 7.8 te/h

Total mass flow out = 99.7 + 7.8 = 107.5 te/h
Concentration of $CaCl_2$ solution = 7.8/107.5 = 0.073 te/te
= 0.073 × 100 = 7.3%
Volume of water out = 99.7 × 1000/1000 = 99.7 m^3/h
∴ Volumetric concentration = 7.8 × 1000/99.7 = 78.2 kg/m^3
= 78.2 g/l.

Ex 8.10 Mass balance on water:

Mass flow at 50°C	5000 kg/h
Mass flow at 5°C	10,000 kg/h
Total mass flow in	15,000 kg/h
∴ Mass flow out	15,000 kg/h

Enthalpy flow at 50°C	5000 × 4.2 × 50 = 1,050,000 kJ/h
Enthalpy flow at 5°C	10,000 × 4.2 × 5 = 210,000 kJ/h
Total enthalpy flow in	1,260,000 kJ/h
∴ Enthalpy flow out	1,260,000 kJ/h

Enthalpy flow out = 15,000 kg/h × 4.2 × T°C
∴ T = 1,260,000/15,000/4.2 = 20°C.

Ex 8.11 The power demand P in kJ/h for an in-line heater to heat 1000 kg/h of water from 10°C to 40°C given by:
1000 × 4.2 × 10 + P = 1000 × 4.2 × 40
∴ P = 126,000 kJ/h = 35 kJ/s or 35 kW. The heater is 90% efficient so power consumed = 35 × 100/90 = 39 kW.

9. Introductory Mass and Heat Transfer

Contents

Aims and Objectives

This unit is an introduction to Mass and Heat Transfer which are important basic concepts for process design. The concepts which you will learn here are important for your understanding of future Modules, assignments and projects.

After studying these notes you should be able to:

1. explain in your own words the following terms and concepts:

   Mass transfer
   Diffusion
   Diffusion coefficient
   Gas and liquid films
   Adsorption coefficient

2. carry out calculations for predicting:

   Adsorption coefficient

It is not essential for you to memorise any of the equations or constants.

It is important that you are able to correctly complete all the self assessment questions at the end of these notes.

Essential Prerequisites

Prior to undertaking this lecture it is important that you have completed and understood the following Units:

Unit 1 Fundamentals of Water Chemistry
Unit 2 Chemical Kinetics and Equilibria
Unit 7 Introduction to Process Engineering
Unit 8 Mass and Heat Balances

9.1 Introduction

Mass Transfer and *Heat Transfer* are two examples of a general physical phenomenon that when there is a *driving force* there will be a flow. In hydraulics the driving force is a pressure difference – or to be precise, a *pressure gradient* which results in a flow of fluid. In electricity a voltage gradient results in a flow of electrical current. In mass transfer the driving force is a concentration gradient causing a flow of mass and in heat transfer the driving force is a temperature gradient and the flow is of heat.

As will become evident, all of these phenomena follow an Ohm's Law type of relationship, that is:

Flow = Driving Force/Resistance

The flow of any quantity arises because of a difference in levels of these quantities in space:

- **Fluid flow** arises from a difference in pressure
- **Heat flow** arises from a difference in temperature
- **Mass flow** arises from a difference in concentration

The flow of all these quantities is subject to resistance imposed by bulk properties of the medium or by elements of the system. Mass flow (or mass transfer) arises from a change in concentration through space, and it therefore stands to reason that this change needs to be defined in order to calculate the rate at which mass transfer takes place.

Section 9.2 requires a study time of about 1 hour

9.2 Mass Transfer Mechanisms

9.2.1 Mass Flow

Mass transfer is one of the most important concepts in water treatment since it controls many of the processes in common use including:

- aeration in activated sludge
- deaeration of boiler feed water
- stripping of carbon dioxide from water
- stripping of volatile organic compounds from water
- adsorption of organics by activated carbon
- removal of dissolved salts by ion exchange

These processes and the related mass transfer mechanisms will be covered in detail in later Modules of the course.

Mass transfer occurs wherever there are two or more components in a mixture and where the components are present at different concentrations (or more accurately 'activities') within the mixture. If the mixture is a single phase then the mass transfer processes are relatively predictable. Heterogeneous systems, however, present phase boundaries at which only empirical information is possible. There are two mechanisms of mass transfer: *convection* and *diffusion*.

9.2.2 Convection

SCHMIDT NUMBER

The Schmidt Number is the ratio of kinematic viscosity to diffusivity

$$Sc = \mu/\rho D$$

The higher the value of Sc the more important is convective mass transfer by comparison with diffusion

Liquid $\approx$ 1000
Gas $\approx$ 1

Convective mass transfer occurs in mixed fluid systems. Mechanical or forced convection will rapidly move the molecules around the system at rates significantly higher than those achieved by diffusion alone. Density differences, caused by a temperature gradient can also cause bulk fluid motion and so accelerate mass transfer. The extent of influence of the convective term can be assessed by comparing the diffusion coefficient with the analogous momentum transfer rate coefficient, the kinematic viscosity $\upsilon = \mu/\rho$. This is represented by the dimensionless group Schmidt number $Sc = \upsilon/D$ (see Section 9.2.7). For a component with D around $10^{-9}\,m^2.s^{-1}$ in water ($\upsilon = 10^{-6}$)$m^2.s^{-1}$, Sc is about 10^3. In contrast, for gases, where diffusion is the dominant transfer mechanism, the value of Sc is approximately unity.

9.2.3 Molecular Diffusion

Ex 9.1

Which of the following is true. The rate of mass transfer is:

1. *Proportional to the diffusion coefficient*
2. *Inversely proportional to the area*
3. *Proportional to the concentration difference*
4. *Inversely proportional to the distance over which the transfer takes place*

The most important mechanism of mass transfer in single phase systems is that of *molecular diffusion*. Molecules vibrate in a random manner moving backwards and forwards at a point. High concentrations of a particular component means that more of its molecules move in one direction than the return. The result is a gradual movement of molecules from high concentrations until a uniform concentration exists throughout the mixture.

Consider a stagnant rectangular volume of the mixture of cross-section A and length l separating concentrations c_1 and c_2 then under steady state conditions:

$$\textbf{Rate of mass transfer by diffusion} = D\,A\,(c_1 - c_2)/l \qquad (9.1)$$

where D = *diffusion coefficient* (a rate constant) or diffusivity ($m^2.\,s^{-1}$)
A = area across which transfer occurs (m^2)
l = distance between the two concentrations

Note that the units for the rate can change according to those of the concentration. c_1 is greater than c_2 so the concentration difference Δc ($c_1 - c_2$) is the *driving force* for transfer and $\Delta c/l$ is called the *concentration gradient*. Values of D vary from large values in the gas phase, through liquids, to very low values in the solid phase (Table 9.1).

Table 9.1 – Diffusion coefficients

System	D, ($m^2.s^{-1}$)
Gases in gases at 273°K	$(0.5–2.0) \times 10^{-5}$
Oxygen in nitrogen	1.81×10^{-5}
Liquids in liquids	$(0.5–4.0) \times 10^{-9}$
CO_2 in water at 20°C	1.77×10^{-9}
NaCl in water at 18°C	1.26×10^{-9}
Ethanol in water at 10°C	0.83×10^{-9}
Oxygen in water at 25°C	2.5×10^{-9}
Solids in solids	$10^{-12}–10^{-34}$

Ex 9.2
Ammonia is absorbed in water from a mixture with air at atmospheric pressure. D for ammonia is $2.3 \times 10^{-5}\,m^2.s^{-1}$. If the distance over which the transfer of mass takes place is 1 mm, what is the mass transfer rate per unit area. The concentration of ammonia is 2.92 mols.m^{-3}

9.2.4 Phase Boundary Limitations

A number of two phase systems are encountered in water treatment:

Liquid – liquid	in solvent extraction
Gas – liquid	in absorption and stripping.
Solid – liquid	in dissolution, leaching, and ion-exchange.
Gas – solid	in fluid-bed catalytic reactors
Gas – liquid – solid	in activated sludge and anaerobic digesters

In spite of considerable fluid motion in the bulk of separate phases that will keep them relatively homogeneous, transfer between phases is controlled by the activities at their interface. At the interface the relative velocity will be zero. Thus mass transfer will be controlled by the problem of moving molecules across a small, perhaps almost stagnant zone of fluid near to the interface. In this zone, the velocity will fall from the bulk value to zero and so also will concentration values change from the bulk to some value at the interface.

Application of Equation 9.1 to this interface is not possible because the zone is not stagnant (there will be some unknown shape of velocity profile) and the zone width is not known. Thus for a cross-sectional area A of this system:

$$\textbf{Rate of mass transfer} = \mathbf{kA\,(c_1 - c_2)} \qquad \mathbf{(9.2)}$$

where k is a *mass transfer coefficient* ($m.s^{-1}$) embodying all that is unknown about the velocity profile width and its effect on molecular diffusion. Mass transfer coefficients must thus be measured for each geometry and fluid flow characteristics of each multi-phase system.

Ex 9.3
Which of the following is true. The rate of mass transfer is:
1. *Proportional to the area of mass transfer*
2. *Inversely proportional to the concentration gradient*
3. *Proportional to the mass transfer coefficient*

9.2.5 Phase Interface

Because of the importance of gas-liquid contact devices both for reaction purposes (eg, oxygen in the activated sludge process) and for gas cleaning (e.g., ammonia absorption), much effort has been directed at understanding the mechanisms of mass transfer across a gas-liquid interface but the concepts are equally applicable to other phase interfaces. An elegant analysis of the interface by Lewis and Whitman (Ind. Eng. Chem. (1924), *16*, 1215) termed the *Two-Film Theory*, gives useful insight. It postulates the existence of two thin films

Ex 9.4
If oxygen is transferred at a rate of 90 mg per minute across a 2 m² boundary where the concentration either side is 2 and 9.5 mg/l, what is the mass transfer coefficient?

at the phase interface (Figure 9.1) and it is within these two films that all mass transfer across the interface takes place. Outside the two films the "bulk" phase is considered to be homogenous.

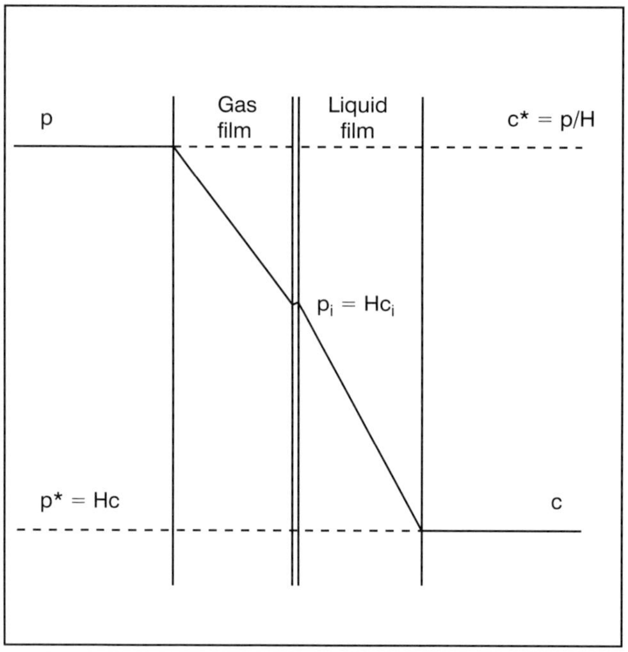

Fig. 9.1 – *Two-film* model of the gas-liquid interface

The rate of mass transfer per unit area (N) can be considered for the:

	Gas film	$\mathbf{N = k_g (p - p_i)}$	**(9.3)**
or	**Liquid film**	$\mathbf{N = k_L (c_i - c)}$	**(9.4)**
or	**Overall**	$\mathbf{N = K (p - c)}$	**(9.5)**

where k_g, k_L and K are gas film, liquid film and overall mass transfer coefficients respectively.

HENRY'S LAW
Henry's Law enables liquid side film concentrations to be calculated from bulk gas pressures and vice versa.
In aeration systems the concentration of oxygen in the liquid at the surface of the bubble can be calculated from the oxygen partial pressure.
The partial pressure of oxygen is 0.2 Atm.

Equation 9.5 contains mixed units in p and c. However, we can define two "equilibrium concentrations" using Henry's Law (Unit 2, 2.4.4):

$c^* = p/H$ is the liquid phase concentration in equilibrium with partial pressure p

$p^* = Hc$ is the partial pressure in the gas phase at equilibrium with a liquid phase concentration c:

Using these equilibrium concentrations we can write:

$$\mathbf{N = K_g (p - p^*)} \quad \textbf{(9.6)}$$

and

$$\mathbf{N = K_L (c^* - c)} \quad \textbf{(9.7)}$$

where K_g and K_L are overall mass transfer coefficients based on gas phase and liquid phase driving forces. If it is assumed that there is no resistance to transfer in the interface and that equilibrium exists, then:

$$\mathbf{p_i = Hc_i}$$

The overall coefficients can now be related to the film coefficients as follows:

$$\frac{1}{K_g} = \frac{p - p^*}{N} = \frac{p - p_i}{N} + \frac{p_i - p^*}{N} = \frac{1}{k_g} + \frac{H(c_i - c)}{N}$$

or

$$\frac{1}{K_g} = \frac{1}{k_g} + \frac{H}{k_L} \qquad (9.8)$$

and similarly

$$\frac{1}{K_L} = \frac{1}{k_L} + \frac{1}{Hk_g} \qquad (9.9)$$

In the case of highly water-soluble gases such as carbon dioxide, ammonia and hydrogen sulphide, H is relatively small, so that the second term on the right hand side of Equation (9.8) can be ignored, giving $K_g = k_g$. This case is referred to as *gas-film controlled*. All the concentration gradient exists in the gas film. For a low solubility gas like oxygen, H is large and the second term on the right hand side of Equation (9.9) is ignored, giving $K_L = k_L$. Thus, the oxygen transfer process is *liquid film controlled*.

FILM RESISTANCE
Transfer of highly soluble gases is gas film controlled.
Transfer of low solubility gases is liquid film controlled (aeration).

9.2.6 Absorption Coefficient

If a volume of liquid is being contacted by a gas stream in a stirred vessel then the total rate of mass transfer, N_T is given by:

$$N_T = k_L A\, \Delta c \qquad (9.10)$$

ADSORPTION COEFFICIENT
The absorption coefficient is defined as

$$k_L a = \Delta c / N$$

Equation 9.10 refers to the whole vessel contents and not just to a portion of surface area, as above. The mass transfer coefficient is an average coefficient for the whole contactor. The area A is the total gas-liquid interfacial area and Δc is an average driving force between the bulk of the liquid and the interface.

The gas transfer rate per unit volume, N, is given by:

$$N = k_L a\, \Delta c \qquad (9.11)$$

The liquid film coefficient is generally considered to be a constant value independent of stirrer speed and air rate. It is the specific interfacial area (a) that changes due to the change in gas hold-up and the mean bubble size. Because 'a' cannot be easily measured in most practical applications, it is combined with k_L to form the *absorption coefficient*, $k_L a$ (h^{-1}).

Ex 9.5
When would you use an absorption coefficient instead of a mass transfer coefficient?

Absorption coefficients are obtained by measurement for all kinds of gas-liquid contact geometries and methods of operation, for example:

- mechanical surface aerators
- diffused air systems
- simple sprays
- packed towers
- sieve tray towers

Ex 9.6
A 200 m^3 atmospheric mixed aeration tank is fitted with bubble diffusers with absorption coefficient 2 h^{-1}. If the influent water has dissolved oxygen concentration 2 mg/l at 20°C, what is the rate of oxygen transfer?

The value of N is divided by Δc to obtain $k_L a$. A suitable value for Δc can be obtained from:

$$\Delta c = c_{Em} - c_m \qquad (9.12)$$

LOG MEAN
The logarithmic mean of x_1 and x_2 is defined as

$$x_{lm} = \frac{(x_1 - x_2)}{\ln\left(\frac{x_1}{x_2}\right)}$$

where c_{Em} is the mean equilibrium concentration at the interface.
c_m is the mean liquid concentration, measured using, eg, a dissolved oxygen electrode.

C_{Em} can be written in terms of the gas phase by $c_{Em} = p_m/H$.

If the gas phase is well mixed then $p_m = p_{out}$ but for plug flow reactors the log mean partial pressure difference must be used:

$$p_{lm} = \frac{(p_{in} - p_{out})}{\ln\left(\frac{p_{in}}{p_{out}}\right)} \qquad \textbf{(9.13)}$$

A value for the absorption coefficient (k_La) enables design calculations to be carried out, by equating the oxygen supply rate to the oxidation rate of the BOD at steady state.

9.2.7 Dimensionless Groups

Ex 9.7
A 200 m³ atmospheric plug flow aeration tank is fitted with bubble diffusers with absorption coefficient 2 h⁻¹. If the influent water has dissolved oxygen concentration 2 mg/l at 20°C, and 50% of the oxygen in the air is dissolved, what is the rate of oxygen transfer?

There are several dimensionless groups (see Unit 7) which are useful in mass transfer and they are defined in Table 9.2.

Table 9.2 – Dimensionless numbers

Dimensionless Group		Representation of:	Definition
Sherwood Number	Sh	$\frac{\text{mass transfer velocity}}{\text{diffusional velocity}}$	$\frac{k.l}{D}$
Schmidt Number	Sc	$\frac{\text{diffusivity of momentum}}{\text{diffusivity of mass}}$	$\frac{\mu}{\rho D}$
Reynolds Number	Re	$\frac{\text{inertial force}}{\text{viscous force}}$	$\frac{\rho.l.v}{\mu}$
Grashof Number	Gr	$\frac{\text{buoyancy force}}{\text{viscous force}}$	$\frac{l^3 g\rho\Delta\rho}{\mu^2}$

Definition of the ratio of various forces through the use of dimensionless groups allows mass transfer in systems to be defined simply as a function of these groups. Such functions can be derived either empirically or from theory, and two of the more important correlations are given below.

For turbulent flow through a pipe:

Ex 9.8
What is the Sherwood number for dissolved CO₂ in water flowing at 2 m/s through a 25 mm pipe at a temperature of 20°C? [NB: refer to Table 9.1 and Unit 7 Table 7.1.]

$$\mathbf{Sh = 0.026.(Re)^{0.8}.(Sc)^{0.3}} \qquad \textbf{(9.14)}$$

For bubbles in a stirred tank:

$$\mathbf{Sh = 0.13.(Gr)^{0.3}(Sc)^{0.3}} \qquad \textbf{(9.15)}$$

These correlations will be covered in more detail in subsequent modules on physical and membrane processes.

9.2.8 Experimental Measurement

Experimental measurement of mass transfer within a system is usually carried out by de-oxygenating the water and then measuring the rate of change of dissolved oxygen concentration (DO) with time. According to Equation 9.10 the rate of change in DO with time is given by:

$$\frac{dC}{dt} = k_L a(C^* - C) \qquad (9.16)$$

where C* is the bulk equilibrium concentration and C the concentration at time t. Integrating the above expression over time period t yields:

$$\ln\left(\frac{C^* - C}{C^* - C_0}\right) = k_L at \qquad (9.17)$$

where C_0 is the original DO concentration (which should then be close to zero). Hence, a plot of the logarithmic term in Equation 9.17 against time yields $k_L a$, i.e. the absorption coefficient, in the slope.

Section 9.3 requires a study time of about 1 hour

9.3 Heat Transfer

9.3.1 Heat Flow

The Second Law of Thermodynamics tells us that where a temperature difference exists between two bodies, heat will flow from the hotter to the colder until a thermal equilibrium is achieved.

In the field of water treatment there are only a few applications for heat transfer, the principal ones being:

- desalination of sea water by distillation
- production of distilled water for pharmaceutical use
- pre-heating of feed water to membrane systems
- indirect heating and cooling systems in industry
- condenser cooling in boiler systems
- heat sanitisation of pharmaceutical water systems
- reactor temperature control for anaerobic digestion

Heat can flow by three mechanisms:

- conduction
- convection
- radiation

All of these are generally considered under heat transfer but, in the context of water treatment, only heat transfer by conduction is usually of importance. The basic principles of heat transfer are considered below but not the detailed design of heat exchangers, which are proprietary items whose design is best left to specialist engineers.

The rate at which heat is transferred along a body by conduction is given by:

JOULES AND CALORIES
There is no defined SI unit of heat. The calorie (cal) is the heat required to raise the temperature of 1 g of water by 1°C and, by inference, the SI unit should be the kcal.

The Joule (J) is the SI unit of work and the relationship between work and heat is:

$$4.2\,\text{J} = 1\,\text{cal}$$

the constant 4.2 is known as the mechanical equivalent of heat.

$$\mathbf{H = -k_t A \frac{dT}{dl}} \qquad \mathbf{(9.18)}$$

where H = heat flow in $J.s^{-1}$
k_t = thermal conductivity $J.s^{-1}.m^{-2}$
A = surface area, m^2
dT/dl = temperature gradient along body, $°C.m^{-1}$

Rearranging we have:

$$\mathbf{H = A.\Delta T/(l/k_t)} \qquad \mathbf{(9.19)}$$

where ΔT = measured temperature difference, °C
l/k_t = *thermal resistance*, R in $°C.m^2/W$

Equation 9.19 is useful because it allows heat transfer through a number of layers of different thicknesses and thermal conductivities to be considered. If there exists a number of layers in series all of the same area, A, and through which the heat flow is H, then:

$$\mathbf{H = A\left(\frac{\Delta T_1}{R_1} + \frac{\Delta T_2}{R_2} + \ldots + \frac{\Delta T_n}{R_n}\right)} \qquad \mathbf{(9.20)}$$

Typical thermal conductivities for common materials in water treatment are given in Table 9.3.

Table 9.3 – Thermal conductivities

	k_t, $W.m^{-1}.°C^{-1}$
Copper	350–400
Aluminium	250–300
Carbon steel	45–60
Stainless steel	15–20
Calcium phosphate scale	3.6
Magnetite	2.9
Calcium carbonate scale	0.8–2.6
Fire brick	1.0
Insulating brick	0.1
Porous silicate scale	0.08–0.14
Rock wool insulation	0.08
Mineral wool insulation	0.03

Consider, for example, the simplest form of heat exchanger comprised of a metal tube with hot water on one side and cold water on the other. Because there is a temperature difference between the two streams, heat will be transferred: the cold water will be heated and the hot water cooled. The situation is analagous to mass transfer in that boundary layers on either side of the tube are envisaged. This arrangement is shown in Figure 9.2 where R_1 is the thermal resistance of the boundary layer on the hot side, R_2 the thermal resistance of the tube itself and R_3 the thermal resistance of the cold boundary layer.

Ex 9.9
A copper tube of thickness 3 mm has an internal layer of calcium carbonate scale 0.5 mm thick. The thermal conductivity of copper is 400 $W.m^{-1}.°C^{-1}$ and for calcium carbonate it is 1.5 $W.m^{-1}.°C^{-1}$. How does the scale layer affect the thermal resistance of the pipe?

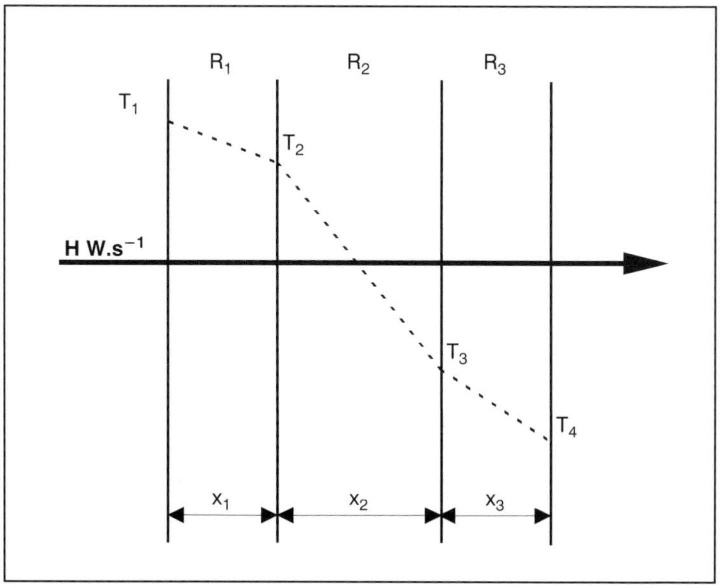

Fig. 9.2 – Heat transfer through multiple layers

Applying a slightly re-arranged Equation 3.1:

For the first layer: $T_1 - T_2 = R_1H/A$
For the second layer: $T_2 - T_3 = R_2H/A$
For the third layer $T_3 - T_4 = R_3H/A$

Adding these gives:

$$T_1 - T_4 = (R_1 + R_2 + R_3).H/A \tag{9.21}$$

If the quantity U, the *overall heat transfer coefficient*, is defined by the reciprocal of the sum of the thermal resistances thus:

$$U = \frac{1}{\sum R_i} \tag{9.22}$$

then the general equation defining heat flow is:

$$H = UA\Delta T \tag{9.23}$$

Clearly the value of U will depend not only on the thermal conductivity of the pipe material but also on the nature of the boundary layers on either side of the heat transfer surface. The most significant characteristic of the boundary layers, as far as water treatment applications are concerned, is whether there is a change of phase from liquid to vapour (*evaporation*) or from vapour to liquid (*condensation*).

The same equation applies where there are layers of different materials, for example a pipe wrapped with lagging or with an internal layer of scale. Heat transfer is such a common unit operation that values of U for many applications have been determined experimentally and are tabulated in many reference books (see bibliography). Some of the most relevant are listed in Table 9.4.

Table 9.4 – Typical heat transfer coefficients

	HOT	COLD	U, $W.m^{-2}.°C^{-1}$
Tube & shell exchanger – steel	water	water	1000–1300
Tube & shell exchanger – steel	steam	water	1300–2100
Plate exchanger – stainless steel	water	water	2000–2500
Jacketed vessel – stainless steel	steam	water	850–1700
Jacketed vessel – glass lined	steam	water	400–570
Unlagged stainless steel pipe	water	air	10–20

Sources: Coulson & Richardson, Perry

9.3.2 Co- and Countercurrent Heat Exchangers

Consider a tubular heat exchanger, that is a tube carrying cold water to be heated with a concentric tube carrying hot water which will do the heating. For there to be heat transfer there must be a temperature gradient all the way along the tube, which means that the temperature on the hot side must be greater than that on the cold side for the whole length of the tube. The temperatures of the two streams will change along the length of the tube so that the temperature difference, ΔT, is not constant and the heat transfer equation is based on the log mean temperature difference, ΔT_{lm} (see Section 9.2.6). Equation 9.23 then becomes:

$$H = UA\Delta T_{lm} \tag{9.24}$$

The temperature profile along the length of the tube will depend on the physical arrangement of the heat exchanger. If the flow in both hot and cold streams is in the same direction – eg from left to right – then the temperature profile will look like that in Figure 9.3. This is called a *co-current* arrangement.

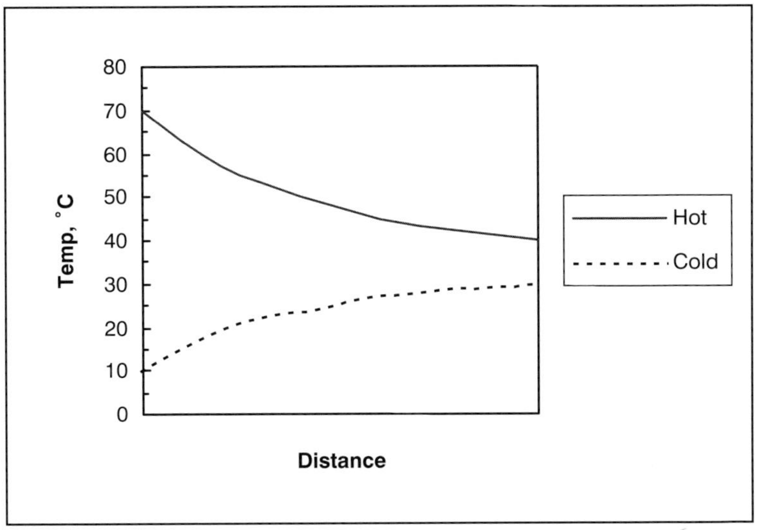

Fig. 9.3 – Temperature profile co-current

If the two streams flow in opposite directions with the hot water flowing from left to right and the hot water from right to left, then the incoming cold water is first exposed to the outgoing hot water and the temperature profile changes to that shown in Figure 9.4, the *countercurrent* arrangement.

As can be seen from the temperature profiles, the mean temperature difference in the countercurrent design is significantly higher than that in the co-current case. This does not affect the quantity of heat transferred but it does affect the *rate* at which the heat is transferred.

Ex 9.10
If the heat transfer coefficient for the exchanger in Figure 9.4 is 1000 $W.s^{-1}.m^{-2}$, calculate the relative heat exchange areas required for:
1. *a co-current heat exchanger*
2. *a countercurrent heat exchanger*

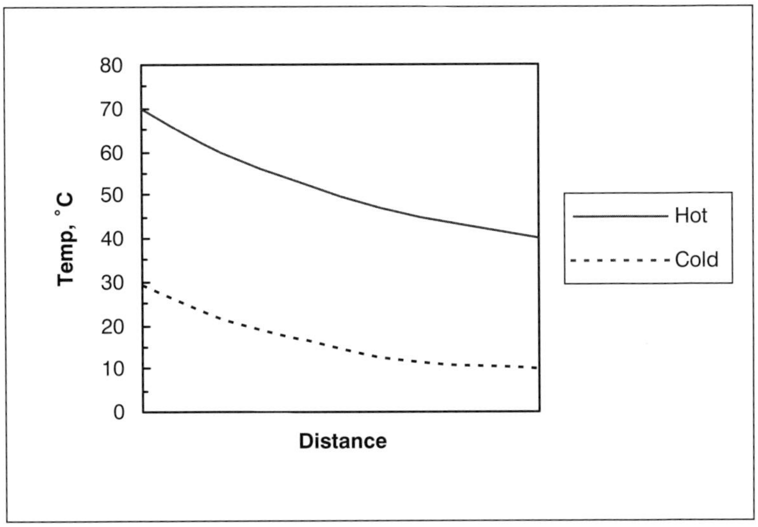

Fig. 9.4 – Temperature profile countercurrent

Suppose that the heat exchanger duty is to heat the water from, say, 10°C to 30°C using hot water at 70°C. This hot water will be cooled to some outlet temperature, T, which will be lower than 70°C and higher than 30°C. The outlet temperature can then be determined by heat balance. For an outlet temperature of 40°C, the system can be represented schematically as shown in Figure 9.5.

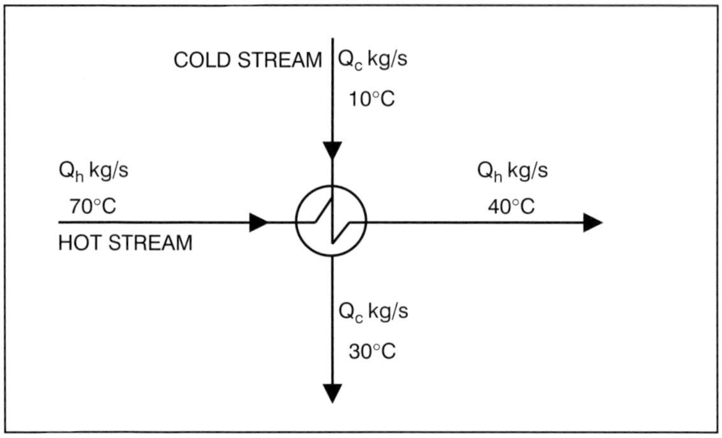

Fig. 9.5 – Schematic representation of heat exchanger

Note that the values of the hot and cold stream flows, Q_h and Q_c, are controlled by only the heat balance and not by the design of the heat exchanger – although some designs might achieve a closer convergence of outlet temperatures than others.

9.4 Self Assessment Questions

SAQ 9.1 An aeration tank operating at 25°C is 2 m deep. At the operating temperature the density of air is 1.43 kg.m^{-3}. Calculate the Sherwood Number and hence the absorption coefficient. [*NB: refer to Table 9.1 and Unit 7 Table 5.1.*]

SAQ 9.2 In a clean water oxygen transfer test at atmospheric pressure and 6°C, a tank containing 18.87 m^3 of water was de-oxygenated by adding sodium sulphite and then aerated by a surface aerator. The dissolved oxygen was measured by a meter in terms of % saturation as follows:

Time (mins)	DO content % saturation
0	11
2	14
4	18
6	22
8	25
10	28.5
12	32
14	35.5
16	38
18	41
20	44
25	50
30	55
35	61

Calculate the k_La value for the surface aerator.

SAQ 9.3 A pharmaceutical water ring main consists of 350 m of 2" diameter 316 L stainless steel tube of wall thickness 0.109". It is intended to sanitise the whole of the ring main once each week by heating to $>$80°C. This will be achieved by heating the circulating water flowing at 13.5 m^3/hr to 90°C at the inlet to the ring main. If the ambient temperature is 18°C what will be the return temperature at the end of the ring main? The overall heat transfer coefficient is 18 W.m^{-2}.°C^{-1}. For the purposes of heat loss calculations assume that the temperature in the ring main is constant.

SAQ 9.4 A stainless steel countercurrent plate heat exchanger is designed to pre-heat 65 m^3/h of reverse osmosis feed water from 10°C to 30°C using a flow of 20 m^3/h of hot water at 80°C. The overall heat transfer coefficient is 2000 W.m^{-2}.°C^{-1}. If a 0.1 mm thick layer of calcium phosphate scale with thermal conductivity 3.6 W.m^{-2}.°C^{-1}, is deposited on the plates what would the hot water inlet temperature need to be increased to in order to maintain the duty?

9.5 Nomenclature

A	area across which mass or heat transfer occurs, m^2
D	diffusion coefficient, $m^2.s^{-1}$
H	Henry's constant
K	overall mass transfer coefficient, $m.s^{-1}$
N	rate of mass transfer flux, $mol.s^{-1}.m^{-2}$
Q	mass flow, $kg.s^{-1}$
R	thermal resistance
U	overall heat transfer coefficient, $W.m^{-2}.°C^{-1}$
T	temperature, °C
ΔT	temperature difference, °C
c	liquid phase concentration, $mol.l^{-1}$
Δc	concentration difference, $mol.l^{-1}$
g	gravitational constant, $m.s^{-2}$
k_g, k_L	gas and liquid film mass transfer coefficients
k_La	adsorption coefficient, h^{-1}
k_t	thermal conductivity, $W.m^{-2}.°C^{-1}$
l	length along which mass or heat transfer occurs, m
p	partial pressure, atm
μ	absolute viscosity, $kg.m^{-1}.s^{-1}$
ν	dynamic viscosity, $m^2.s^{-1}$
ρ	density, $kg.m^{-3}$
Gr	Grashof Number
Re	Reynolds Number
Sc	Schmidt Number
Sh	Sherwood Number

9.6 Solutions to Exercises

Ex 9.1 Answers 1, 3, 4 are true, see equation 9.1

Ex 9.2 Assume initial concentration in water is zero ($c_2 = 0\,mols.m^{-3}$) therefore $\Delta C = C_1 - C_2 = 2.92 - 0 = 2.92\,mols.m^{-3}$
$N/A = D(\Delta C)/L = (2.3 \times 10^{-5} \times 2.92)/(1 \times 10^{-3}) = 0.067\,mol.m^{-2}.s^{-1}$

Ex 9.3 Answers 1, 3 are correct, see equation 9.2, compare to answer to Ex 9.1.

Ex 9.4 Rate is 90 mg in 1 min (60 s) therefore rate $= 90/60 = 1.5\,mg.s^{-1}$
rate $= k \times$ area $\times \Delta c$
$1.5\,mg/s = k \times 2\,m^2 \times (9.5-2) \times 1000\,mg/m^3$
So, $k = 1.5/(2 \times 7.5 \times 1000) = 1 \times 10^{-4}\,m/s$

Ex 9.5 Absorption coefficients are used when the area of transfer is unkown.

Ex 9.6 Total pressure, P = 1 atm
Partial pressure of oxygen (p_{O2}) $= 0.2\,P = 0.2$
Henrys constant for O_2 at 20°C (Unit C2) $= 0.024\,atm.l.mg^{-1}$
$C_{Em} = p_{O2}/H = 0.2/0.024 = 8.3\,mg/l = 8.3\,g.m^{-3}$
$c_m = 2\,mg/l = 2\,g.m^{-3}$
$k_La = 2\,h^{-1}$
$N = 2 \times (8.3-2) = 12.6\,g.h^{-1}m^{-3}$
$V = 200\,m^3$
$N_T = 12.6 \times 200 = 2520\,g.h^{-1} = 2.5\,kg/h$

Ex 9.7 The calculation is similar to that above but instead of taking the partial pressure of oxygen as 0.2 P, the log mean partial pressure is used. The inlet partial pressure is 0.2 and the outlet (50% of the oxygen being absorbed) is 0.1 so from Eqn 9.13:

$p_{lm} = (0.2-0.1)/\ln(0.2/0.1) = 0.1/0.69 = 0.14$

Now
$c_0 = 2\,mg/l = 2\,g.m^{-3}$
$c_e = 0.14/0.024 = 5.8\,mg/l = 5.8\,g.m^{-3}$
$k_La = 2\,h^{-1}$
$N = 2 \times (5.8-2) = 7.7\,g.h^{-1}.m^{-3}$
$V = 200\,m^3$
$N_T = 7.7 \times 200 = 1530\,g.h^{-1} = 1.5\,kg/h$

Ex 9.8 At 20°C, density $= 1000\,kg/m^3$, viscosity $= 0.001\,kg\,m^{-1}s^{-1}$,
From Table 2.1, $D = 1.77 \times 10^{-9}\,m^2/s$.
$Sh = 0.026\,Re^{0.8}\,Sc^{0.3}$
$= 0.026\,(10^6 \times 2 \times 0.025)^{0.8}\,(10^{-6}/1.77 \times 10^{-9})^{0.3}$
$= 0.026 \times 5740 \times 6.69$
$= 999$

Ex 9.9 Thermal resistance for pipe $= x/k_t = 0.003/400$
$= 7.5 \times 10^{-6}\,m^2.°C.W^{-1}$
Thermal resistance for scale $= 0.0005/1.5 = 3.3 \times 10^{-3}$
Total thermal resistance R given by
$R = 7.5 \times 10^{-6} + 3.3 \times 10^{-3} = 3.3 \times 10^{-3}$
R is increased by $3.3 \times 10^{-3}/7.5 \times 10^{-6} = 440$ times.

Ex 9.10 Heat flow, H, in each case is the same

$U = 1000\,W.s^{-1}.m^{-2}$

$\therefore H = UA\Delta T_m$

$\therefore A = H/1000/\Delta T_m$

Co current $\Delta T_{m:}$

$\Delta T_{in} = 70 - 10 = 60°C$

$\Delta T_{out} = 40 - 30 = 10°C$

$\Delta T_m = (60 - 10)/\ln(60/10) = 50/\ln 6 = 27.9°C$

$A_1 = H/27{,}900$

Countercurrent

$\Delta T_{in} = 70 - 30 = 40°C$

$\Delta T_{out} = 40 - 10 = 30°C$

$\Delta T_m = (40 - 30)/\ln(40/30) = 10/\ln 1.3 = 34.8°C$

$A_2 = H/34{,}800$

$A_1/A_2 = 34{,}800/27{,}900 = 1.25$ so area for co-current exchanger is 25% larger than for countercurrent.

10. Reactor Design Theory

Contents

Aims and Objectives

This Unit covers the fundamentals underlying the design of chemical reactors.

After studying these notes you should be able to:

1. explain in your own words the following terms and concepts:

   reactor mass balance
   steady-state mass balance
   first order rate reaction
   batch reactors
   continuous reactors
   continuous stirred tank reactor (CSTR)
   plug-flow tank reactor (PFTR)
   residence time

2. identify the operation and performance characteristics of batch, CSTR and PFTR processes

3. carry out a reactor mass balances for batch, CSTR and PFTR processes

4. apply the appropriate ideal reactor equations to enable the determination of final reactant concentration, reactor size, reaction time and residence time.

It is not essential for you to memorise any equations or constants.

It is important that you are able to correctly complete all the self assessment questions at the end of these notes.

Essential Prerequisites

Prior to undertaking this Unit it is important that you have completed and understood the following Units:

Unit 2 Chemical Kinetics and Equilibria
Unit 4 Fundamentals of Microbiology
Unit 7 Fundamentals of Process Engineering
Unit 8 Mass and Heat Balances

10.1 Introduction

Any item of plant in which a chemical or biological reaction or a physical separation process takes place is called a *reactor*. In engineering parlance they are often, identified by the type of reaction so, commonly, you will hear *oxidation tank*, *ion exchanger*, *filter*, *clarifier*, *adsorber*, and so on but they are, nevertheless, all types of reactor.

Both batch and continuous processes are used in water and wastewater treatment and it is important to understand the differences. The mode of operation has a significant effect on:

- the rate of the reaction
- the substrate concentration at a given time
- the reactor size required

In *batch* processes materials are added to the reactor at the start, the system is then closed and product removed only when the process is complete. The reaction occurs with no inflow to and outflow from the reactor. The reactor is, ideally, completely mixed or *homogeneous* – that is the concentrations of all components are the same at all points in the reactor.

In *continuous* processes materials flow to and from the reactor during the reaction. *Continuous stirred tank reactors* (CSTRs) aim to be completely mixed so that the contents are homogeneous while *plug flow tank reactors* (PFTRs) aim to avoid mixing so that concentrations and reaction conditions change along the length of the reactor.

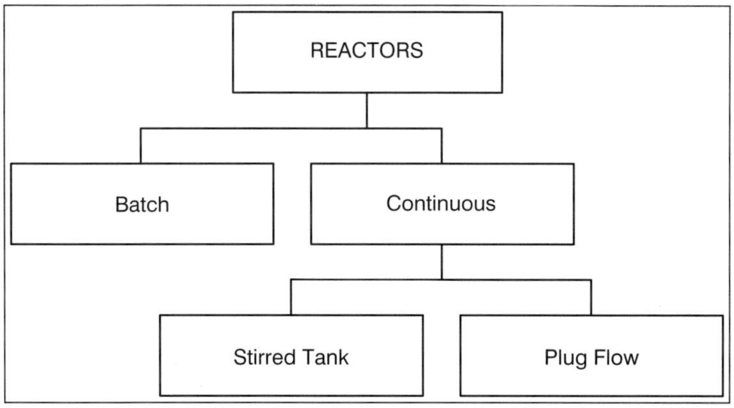

Fig. 10.1 – Reactor types

This Unit will look at the basic theory of batch, CSTR and PFTR reactors and at how the behaviour of real processes can be assessed in terms of the ideal models. Such models can be used to determine the size of the reactor required, the time required for the reaction and the pollutant effluent concentration attained.

10.2 Mass Balance

10.2.1 A General Expression

MASS BALANCE
Under steady state conditions the total masses entering and leaving a reactor are equal.

In Unit E2 we noted that, under steady state conditions, the total mass entering a reactor will be equal to the total mass leaving. We will use this basic steady-state mass balance as a first step in developing an idealised reactor model:

$$\textbf{In} = \textbf{Out} + \textbf{Removal} + \textbf{Accumulation} \qquad \textbf{(10.1)}$$

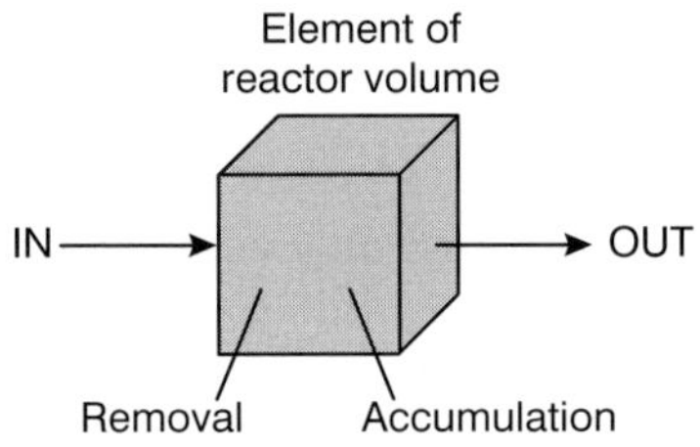

Fig. 10.2 – Steady state mass balance

FIRST ORDER REACTION
The relationship between reaction rate and substrate concentration is:
$-r = k.C$

In water and wastewater treatment processes the removal of a pollutant can be described by numerous reaction rates. In this Unit only one type of reaction, namely, an homogenous *first order reaction*, is considered. From Unit 2 a first order reaction is represented by the equation:

$$-\mathbf{r} = \mathbf{kC} \qquad \textbf{(10.2)}$$

where r = rate of reaction, $mol.l^{-1}.s^{-1}$
C = concentration of reactant, $mol.l^{-1}$
k = first order rate constant, s^{-1}

For many treatment systems, such as disinfection, biological degradation, chemical oxidation and flocculation, where the pollutant is being removed at a rate proportional to the concentration of pollutant remaining, assuming a first order reaction is reasonable.

Ex 10.1
The concentration of solids in a sand filter is 25 mg. l^{-1}. The filter has a reaction rate constant of 0.002 s^{-1}. What is the volumetric rate of reaction?

The rate of reaction is defined as the rate of change in concentration of a reactant or product. A reactor of volume V, contains a substance A at an initial concentration C_0. At time t the concentration of A falls to C. The number of moles of A in volume V is M. Hence, C = M/V. Since r is the number of moles reacting per unit volume per unit time (the *volumetric rate of reaction)*, it follows that the rate of fall in concentration of A can be written in a differential form:

$$-\mathbf{r} = \frac{\mathbf{1}}{\mathbf{V}}\left(\frac{\mathbf{dM}}{\mathbf{dt}}\right) = \frac{\mathbf{dC}}{\mathbf{dt}} = \mathbf{kC} \qquad \textbf{(10.3)}$$

where V = reactor volume
M = number of moles of A
t = reaction time

The implications of conducting such a reaction in the three types of idealised reactor – batch, PFTR and CSTR – may now be considered.

10.3 Idealised Reactors

Section 10.3 requires a study time of about 3 hours

10.3.1 Batch Reactors

Some processes are operated under batch conditions where the reactants are entered and then the reaction allowed to proceed until the desired conversion has been achieved. There are no inflows or outflows from the unit and so no "product" is available until the reaction is complete. To overcome the stop-start nature of product formation, multiple batch reactors are often operated in a cyclic manner to achieve semi-continuous operation.

Ex 10.2
Which of the following best describes a first-order rate reaction?
1. *A reaction rate that is independent of substrate concentration*
2. *A reaction rate that is dependent on reaction time*
3. *A reaction rate that is dependent upon substrate concentration*

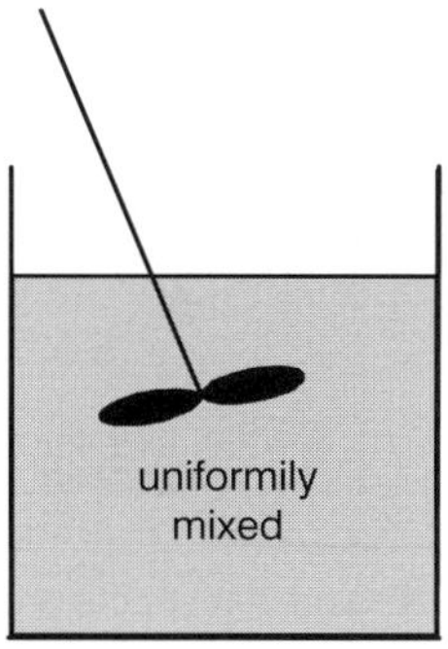

Fig. 10.3 – Batch reactor

Batch reactors are commonly small scale and are suitable for reactions that proceed with a low rate of reaction, hence can be left for longer periods of time rather than needing a larger reactor, as is the case with continuous reactors, as we shall see later.

BATCH REACTORS
Batch reactors are *closed systems*; feed is added at the beginning of the process and products removed at the end.

Carrying out a mass balance (Equation 10.1) of a *batch reactor* gives:

$$\text{In} = 0$$
$$\text{Out} = 0$$
$$\text{Removal} = (-r)V$$
$$\text{Accumulation} = -v\frac{dC}{dt}$$

Hence:

$$(-r) = \frac{-dC}{dt} \qquad (10.4)$$

Substituting for r from the rate equation for a first order reaction (Equation 10.3):

$$kC = \frac{-dC}{dt} \qquad (10.5)$$

The solution to this differential equation is the design equation for a batch reactor:

$$-kt = \ln\left(\frac{C_F}{C_0}\right) \qquad (10.6)$$

or

$$C_F = C_0 e^{(-kt)} \qquad (10.7)$$

where: C_0 = initial concentration (at time t = 0)

Ex 10.3
Which of the following is true of batch reactors?
1. *Multiple batch reactor processes enable continuous operation.*
2. *Batch reactor operation is labour intensive.*
3. *Product concentration changes with reaction time.*
4. *Product concentration changes with position within reactor.*
5. *Batch reactor operation is very similar to continuous reactor operation.*

C_F = final concentration (at time t = t)
t = time of operation

Equation 10.7 is the first order rate reaction in an exponential form. If any three of the four variables are known the fourth can be calculated.

Ex 10.4
The destruction of coliform bacteria by chlorine is a first order reaction with rate constant, $k = 0.089\,min^{-1}$.

Water containing 100 cfu/ml is to be disinfected so that the final coliform count is less than 0.1 cfu/ml. What contact time between chlorine and water is required to effect this reduction?

The design of real reactors is often carried out graphically as the exact nature of how the rate of reaction change is determined experimentally. If the reciprocal of the reaction rate is plotted against the concentration, C, then this depicts Equation 10.5 and the area under the curve is equal to the time required for the batch reaction to proceed to the desired concentration.

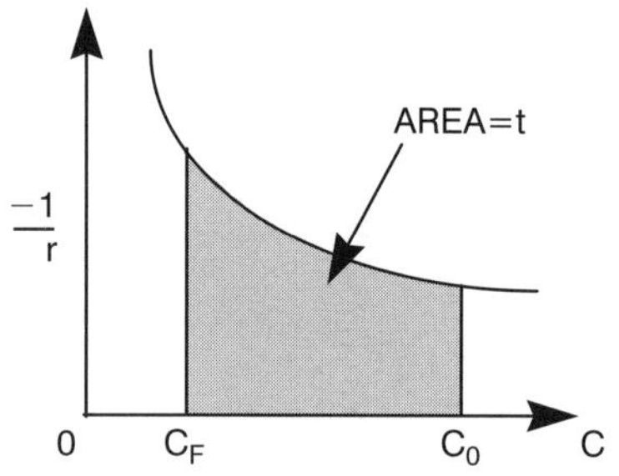

C_F is the concentration of the reactant at the end of the batch operation.

Fig. 10.4 – Graphical representation of equation 10.5

Ex 10.5
Assuming no lag time and 1st order kinetics when is reaction rate highest during the operation of a biological batch reactor?
1. *at the start of the process*
2. *at the middle of the process*
3. *at the end of the process*

In the equations presented for batch processes the time variable, t, is the time required for the reaction process to occur. In reality, operating batch reactors involves long periods of downtime; unproductive periods when the reactor is being filled, emptied and cleaned. Furthermore, in biological processes, a lag time occurs at the start of the process during which there is either no or only very slow removal of pollutants.

10.3.2 Continuous Reactors

In water and wastewater treatment it is usually the case that a continuous flow of treated water is required or that there is a continuous discharge of wastewater from a process which has to be treated. It is clearly advantageous to run treatment processes continuously such that the reaction occurs whilst material flows into and out of the reactor. In such processes time is not a variable as the system is operated under steady state conditions. Instead a concept known as residence time is introduced which describes the time that material remains in the reactor.

$$\tau = \frac{\textbf{volume}}{\textbf{flow}} = \frac{\mathbf{V}}{\mathbf{Q}} \qquad \textbf{(10.8)}$$

where τ = residence time, h
V = reactor volume, m^3
Q = process flow rate, $m^3.h^{-1}$

Hence, if the residence time of a continuous process were, say, 2 hours then one *reactor volume* of feed would be processed every 2 hours.

Whilst continuous processes are effective for relatively fast reaction rate processes and allow high levels of automatic control, they need more supporting equipment and are generally higher in capital cost.

Continuous reactors are operated in one of two modes: continuously stirred tank reactors (CSTR) or plug-flow tank reactors (PFTR).

10.3.3 Stirred Tank Reactors

In the continuous stirred tank reactor (CSTR) all the components within the reactor vessel are completely mixed during operation. As the process is operated under completely mixed conditions the concentration of the components in the reactor is the same as the concentration of the components in the outlet (C_F).

CSTR
The effluent from an ideal Continuous Stirred Tank Reactor has the same composition as the liquid within the reactor. Substrate entering a CSTR is immediately diluted to the outlet concentration.

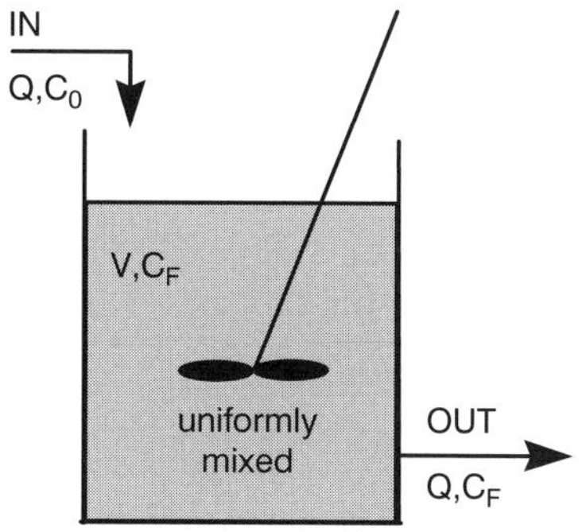

Fig. 10.5 – CSTR

The CSTR is common in the water and wastewater treatment industries and are relatively simple in design and operation. The activated sludge process for sewage treatment, neutralisation of acidic solutions and coagulation of colloids by chemical addition are typical examples in water treatment.

Carrying out a mass balance (Equation 10.1) over a *CSTR* gives:

$$\begin{aligned} \text{In} &= QC_0 \\ \text{Out} &= QC_F \\ \text{Removal} &= (-r)V \\ \text{Accumulation} &= 0 \end{aligned}$$

Hence:

$$\mathbf{QC_0 = QC_F + (-r)V} \tag{10.9}$$

that is:

$$\mathbf{Q(C_0 - C_F) = (-r)V} \tag{10.10}$$

As the concentration, C_F, is uniform throughout the reactor, i.e. it is homogeneous, Equation 10.10 can be rearranged to:

$$\mathbf{\frac{(C_0 - C_F)}{(-r)} = \frac{V}{Q}} \tag{10.11}$$

Substituting from Equation 10.8 gives:

$$\tau = \mathbf{\frac{C_0 - C_F}{(-r)}} \tag{10.12}$$

Substituting the rate equation for a first order reaction (Equation 10.3) yields the following expression for a CSTR:

$$k\tau = \frac{C_0 - C_F}{C_F} \qquad \textbf{(10.13)}$$

Equation 10.13 can be used to calculate the size of reactor required to achieve a desired outlet concentration, C_F. Alternatively the outlet concentration attained from a known reactor volume can be determined. This problem can also be carried out graphically, as shown in Figure 10.7. The concentration in a CSTR is constant at the reactor outlet value (C_F) and the residence time is equal to the area of the rectangle formed at the point where the inverse rate curve meets the vertical drawn from the outlet concentration on the x-axis.

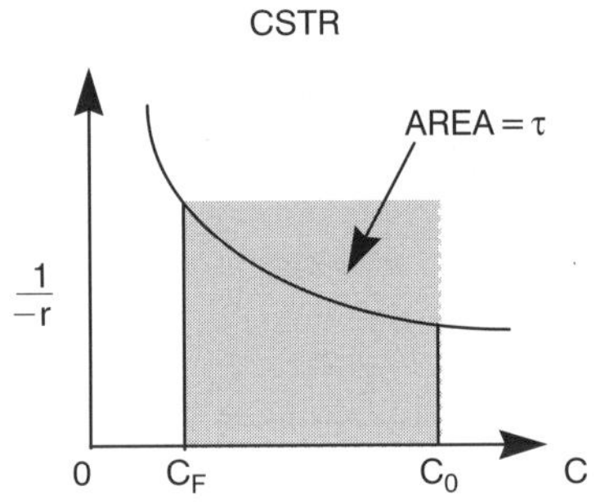

Fig. 10.6 – Graphical estimation of reaction time

10.3.4 Plug Flow Reactors

> **PLUG FLOW REACTOR**
> Liquid flow in a PFTR has a constant velocity with all elements having identical residence times. No mixing occurs in an ideal PFTR. In a PFTR treating wastewater the pollutant concentration will high at the influent end and low at the effluent end.

In a plug-flow tank reactors (PFTR) the feed enters the reactor and, whilst passing along its length, is converted into product. An ideal PFTR has no longitudinal mixing so an *element* of fluid does not interact with neighbouring elements in the reactor.

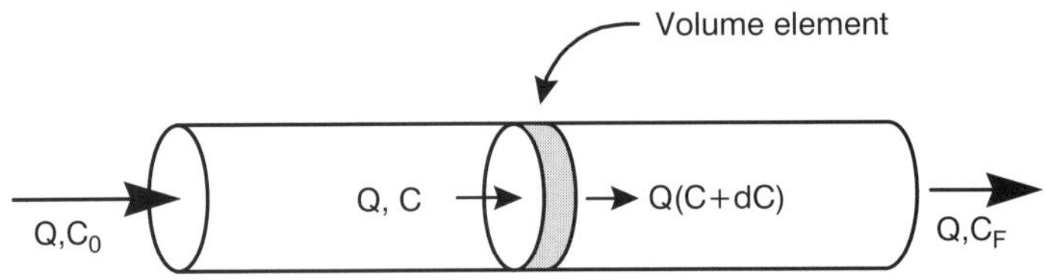

Fig. 10.7 – Plug flow reactor

Ex 10.6
Which of the following best describes a PFTR removing trace organics from water for domestic use?
1. *During operation no untreated water is added to the system.*
2. *The concentration of organics within the reactor changes with time.*
3. *The concentration of organics changes along the length of the reactor.*

A PFTR can be visualised as a number of interconnected batch reactors each of the volume of the element. The component concentration, C, changes along the length of the reactor just as the concentration in a batch reactor changes with time, the distance along the reactor length being a direct measure of the residence time.

The design equation can be derived by conducting a mass balance. However, as the concentration changes with position the mass balance must be split up such that the reactor is divided into many individual elemental reactors where the concentration change is only small. These individual cells can then be added together to determine the overall effect. This is achieved mathematically by integrating the mass balance of an individual element between the limits at the two ends of the reactor.

Carrying out a mass balance along an element in a *PFTR* gives:

$$\begin{aligned} \text{In} &= QC_0 \\ \text{Out} &= Q(C + dC) \\ \text{Removal} &= (-r)dV \\ \text{Accumulation} &= 0 \end{aligned}$$

hence,

$$\mathbf{QC_0 = Q(C_0 + dC) + (-r)dV} \qquad \textbf{(10.14)}$$

which simplifies to the differential equation:

$$\mathbf{\frac{dC}{dV} = \frac{r}{Q}} \qquad \textbf{(10.15)}$$

If r is substituted from the rate equation for a first order reaction (Equation 10.3) we get:

$$\mathbf{\frac{kC}{Q} = \frac{-dC}{dV}} \qquad \textbf{(10.16)}$$

The solution to this differential equation is the design equation for a plug flow reactor:

$$\mathbf{\ln C_F - \ln C_0 = \frac{-kV}{Q}} \qquad \textbf{(10.17)}$$

If τ is then substituted from Equation 10.8 the above equation becomes:

$$\mathbf{-k\tau = \ln\left(\frac{C_F}{C_0}\right)} \qquad \textbf{(10.18)}$$

or

$$\mathbf{C_F = C_0 e^{(-k\tau)}} \qquad \textbf{(10.19)}$$

The graphical representation of a PFTR reactor (Figure 10.8) is similar that for a batch reactor; in this case the area under the graph is equal to the residence time and hence reactor volume can be derived.

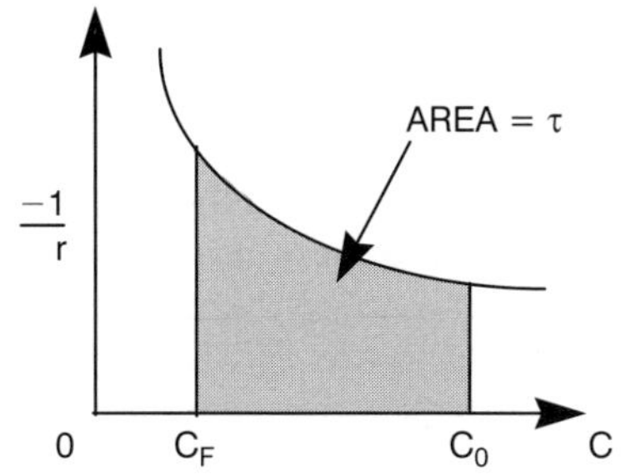

Fig. 10.8 – Graphical representation of plug flow reactor

Ex 10.7
In a wastewater treatment process removing phenolic pollutants the pollutant concentration is uniform throughout the reactor and equal to the effluent concentration. Which type of idealised reactor(s) are being described?
1. *a PFTR*
2. *a batch reactor at the end of operation*
3. *a CSTR*

10.3.5 Comparison of Idealised Reactors

In water and wastewater treatment the majority of the processes are continuous; batch processes are impractical due to the large volumes involved in water purification and wastewater remediation.

The simplest way to choose which mode of continuous operation, either CSTR or PFTR, to opt for is by comparing the residence times for both systems for the treatment of a wastewater to the same standard. In this case Q, C_0 and C_F are the same for each system.

From Equations 10.13 and 10.18 the residence time for a CSTR and a PFTR are respectively given by:

$$\tau_{CSTR} = \frac{1}{k}\left[\frac{C_0}{C_F} - 1\right] \quad \textbf{(10.20)}$$

and

$$\tau_{PFTR} = \frac{1}{k}\ln\left(\frac{C_0}{C_F}\right) \quad \textbf{(10.21)}$$

As the flowrate, Q, is the same in both the CSTR and PFTR, the ratio of the reactor volumes is equal to the ratio of the residence times:

$$\frac{\tau_{CSTR}}{\tau_{PFTR}} = \frac{V_{CSTR}}{V_{PFTR}} = \frac{\left[\frac{C_0}{C_F} - 1\right]}{\ln\left(\frac{C_0}{C_F}\right)} \quad \textbf{(10.22)}$$

Graphically this would be the ratio of the shaded areas from the two graphs in Figure 10.9:

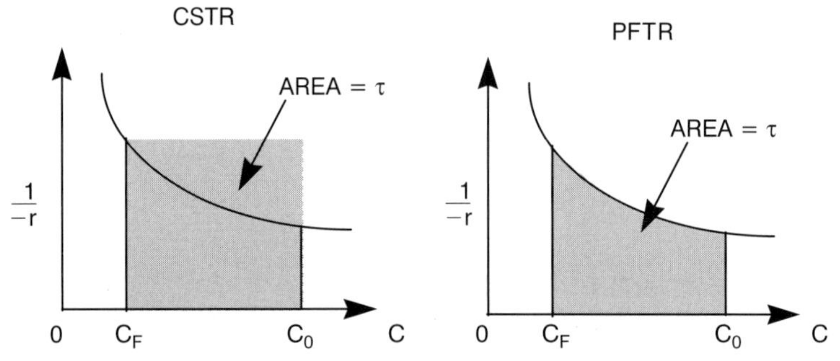

Fig. 10.9 – Residence times for CSTR and PFTR

POLLUTANT REMOVAL
Pollutants entering a CSTR are immediately diluted to the outlet concentration. In 1st order and Michealis-Menten reactions, where the rate of reaction is proportional to the substrate concentration, CSTRs achieve lower rates of pollutant removal than PFTRs and batch reactors of the same volume.

The concentration in the CSTR is the same as the outlet concentration (C_F). In the PFTR the concentration only falls to the outlet concentration at the outlet and remains above this concentration throughout the rest of the reactor. As the rate of reaction is proportional to concentration in 1st order reactions, the reaction proceeds more quickly in the PFTR and hence the reactor volume required is smaller.

To illustrate this if it is assumed that a 90% treatment is required then the ratio of C_0/C_F is 10. Substituting this into Equation 10.22 results in a ratio of residence times of 3.91. In other words, a CSTR would be almost four times the size of a plug-flow reactor to achieve the same degree of treatment.

However, in practice the choice between PFTR and CSTR is not based just on which mode gives the fastest rate of reaction, and hence the smallest reactor volume. Ideal PFTR conditions are harder to achieve and maintain compared to ideal CSTR conditions. Most PFTRs used in water treatment are of the *packed bed* type in which water is contacted with some sort of granular media (typical applications include ion exchange, adsorption processes, sand filtration and fixed film biological processes). The flow conditions are far removed from the idealised tubular plug flow reactor considered in Section 10.3.4. Even in plug flow activated sludge plants, the cross sectional area of the *aeration lane* is sufficiently large that back-mixing, channelling and velocity fluctuations occur. Also, in water and wastewater treatment plants, influent composition varies seasonally or even diurnally. Because of the rapid dilution of the influent stream in a CSTR these variations have less impact than they would in a PFTR.

Table 10.1 compares the characteristics of CSTR and PFTR for the same flow and concentration duty.

Table 10.1 – Comparison of CSTR and PFTR

	CSTR	PFTR
Volume	larger (typically 4x)	smaller
Footprint	larger	smaller
Capital cost	usually higher	usually lower
Operating cost	higher (mixing power)	Lower
Flow control	simple	difficult
Maintenance	generally lower	generally higher
Effect of feed change	little impact	possible high impact

Ex 10.8
Which of the following best describes a PFTR treating an industrial wastewater by employing a support medium for biofilm growth?
1. *The support medium helps prevent bacteria from continuously being washed out of the system.*
2. *The support medium increases the volume available for the biodegradation of pollutants in the wastewater.*
3. *Fluctuations in the concentration of pollutants entering the treatment process are diluted in the reactor due to the mixing of liquid within the reactor.*

10.3.6 Stirred Tank Reactors in Series

The performance of an ideal PFTR can be approached by connecting multiple smaller CSTRs in series, which avoids some of the difficulties associated with PFTR operation.

Figure 10.10 shows a series of equal-sized CSTRs. The concentration entering each reactor is equal to the concentration in the previous reactor; thus the concentration is reduced in a step-wise manner as opposed to the instant reduction that would occur in a single large CSTR.

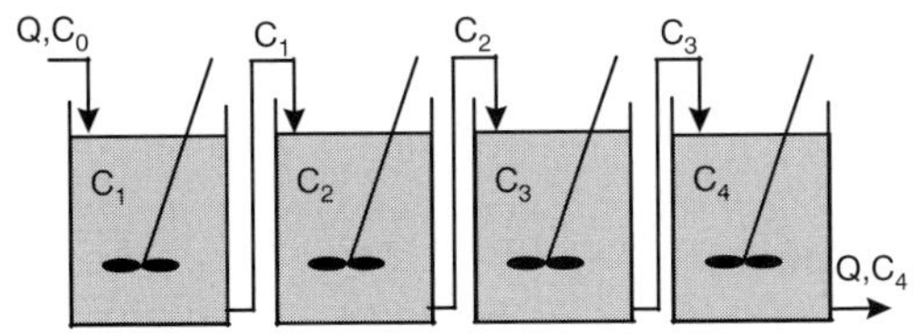

Fig. 10.10 – CSTR in series

The benefit of this results from the fact that the rate of reaction is proportional to the concentration in first order reactions. This means that in all but the final tank the rate of reaction will be greater than would occur in a single tank. Consider applying the single CSTR Equation 10.15, to any reactor in the train – say the ith. The inlet concentration to the ith reactor is the concentration in the previous tank $C_{(i-1)}$ and the outlet concentration is C_i; thus:

$$\frac{C_{(i-1)}}{C_i} = (1 + k\tau_i) \qquad (10.23)$$

This can be extended for N reactors in series to determine the overall performance by multiplying individual reactor performances to form a mathematical series which can be simplified to:

$$\frac{C_0}{C_j} = \frac{C_0}{C_1}.\frac{C_1}{C_2}\cdots\frac{C_{N-1}}{C_N} = (1 + k\tau_1).(1 + k\tau_2)\cdots(1 + k\tau_N) \quad (10.24)$$

that is:

$$\frac{C_0}{C_N} = (1 + k\tau)^N \quad (10.25)$$

This result can be illustrated graphically and compared to a single CSTR and a PFTR to highlight the benefits of using multiple CSTR reactors in series.

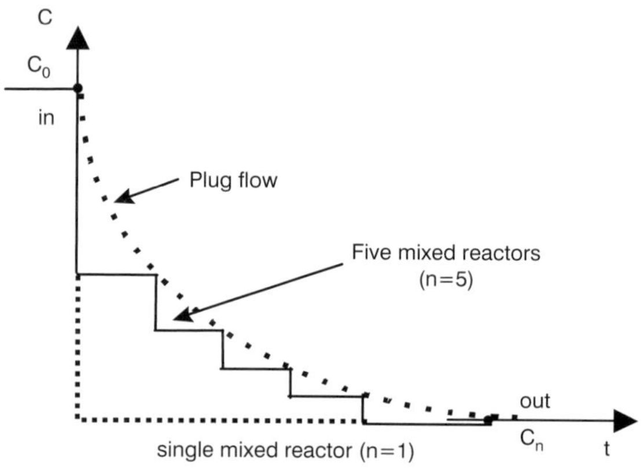

Fig. 10.11 – Comparison of multiple CSTR and PFTR

Ex 10.9
Based on the same k value, calculate the ratio of the reactor volumes required using:
a. a single CSTR
b. six CSTRs in series
c. a single PFTR
for 99% removal.

An example of such a system is the tapered aeration plant employed at Cotton Valley wastewater treatment plant at Milton Keynes where four CSTR reactors are connected in series to maintain a higher concentration of pollutants at the front end of the treatment process.

As the number of CSTRs in series is increased, the behaviour of the plant becomes more like that of a PFTR which is, effectively, an infinite number of CSTRs in series. The step-wise reductions in the line that describes CSTRs in series would become so small as to be the same as the curve that describes a single PFTR.

10.3.7 Fed Batch Reactors

Another variation which is found mainly in biological wastewater treatment is the *fed batch reactor*. This is, effectively, an automated batch reactor adapted for semi-continuous processing. The reactor is filled with reactant, processed and then the products are drawn out, hence the reactor is sometimes called a *fill and draw* reactor. Under these conditions, the reactor operates like the conventional batch reactor, analysed in Section 10.3.1. However, usually the reactor is not completely emptied at the end of the batch and some of the products from the last batch are deliberately left behind to dilute the influent. This means that it exhibits some of the characteristics of a CSTR.

Often wastewater treatment is carried out in a number of steps. They can be carried out in a single reactor as a *sequence* of steps rather than a single reaction. A typical wastewater treatment sequence might consist of the following steps:

1. fill the reactor
2. anoxic mixing for 2 hours to achieve denitrification
3. mixing with aeration for 16 hours for BOD reduction and nitrification
4. settlement for 2 hours to clarify the treated wastewater
5. decantation to discharge the treated wastewater
6. sludge removal

A reactor designed to carry out these steps automatically is often called a *sequencing batch reactor* (SBR) and has a number of advantages over continuous reactors in certain wastewater applications, as will be shown in a later Module. The drawback to the SBR is that to handle a continuous flow a number of reactors (typically at least four) operating in parallel are required, each "out of phase" with the others.

10.4 Reactor Design Worked Example

10.4.1 Problem

A pilot-plant CSTR and PFTR are undergoing trials at a municipal wastewater treatment works. The reactors each have a reactor volume of $2\,m^3$, and both are treating the same wastewater at a flow rate of $5 \times 10^{-3} m^3 s^{-1}$ and a pollutant concentration of 500 mole m^{-3}. A 1st order reaction with a rate constant of $1.5 \times 10^{-3}\ s^{-1}$ is assumed for both reactors.

Calculate the removal efficiency, E, achieved in both the CSTR and PFTR pilot-plants.

10.4.2 Solution

Data $C_0 = 500$ mole m^{-3}
$k = 1.5 \times 10^{-3}\ s^{-1}$
$V = 2\,m^3$
$Q = 5 \times 10^{-3}\ m^3 s^{-1}$

The equation for removal efficiency is:

$$\mathbf{E = \frac{C_0 - C_F}{C_0}} \quad \mathbf{(10.26)}$$

Therefore, to determine E the outlet concentration, C_F, first needs to be calculated. For a CSTR Equation 10.15 can be used to calculate C_F:

$$\frac{C_0}{C_F} = 1 + k\tau$$

However, the residence time, τ, needs to first be calculated from Equation 10.9:

$$\tau = \frac{V}{Q} = \frac{2}{5 \times 10^{-3}} = 400\,s$$

Substituting into Equation 10.15:

$$\frac{C_0}{C_F} = 1 + (1.5 \times 10^{-3}) \cdot (400)$$

$$\frac{C_0}{C_F} = 1.6$$

$$C_F = \frac{C_0}{1.6} = \frac{500}{1.6}$$

$$C_F = 312.5$$

Substituting into Equation 10.26 gives:

$$E = \frac{500 - 312.5}{500}$$

$$E = 0.375$$

This indicates that the pilot-plant CSTR is capable of 37.5% removal efficiency.

For a PFTR Equation 10.22 can be used to calculate C_F:

$$\mathbf{C_F = C_0 e^{(-k\tau)}}$$

As the PFTR is treating the same flow, Q, and has the same volume, V, as the CSTR the residence time is also the same. Thus, substituting into Equation 10.22

$$\mathbf{C_F = 500.e^{(-(1.5\times10^{-3}).(400))}}$$

$$\mathbf{C_F = 274.4}$$

Substituting into Equation 10.26:

$$\mathbf{E = \frac{500 - 274.4}{500}}$$

$$\mathbf{E = 0.45}$$

This indicates that the pilot-plant PFTR is capable of 45% removal efficiency.

Comparison of reactors types:

$$\mathbf{\frac{E_{PFTR}}{E_{CSTR}} = \frac{0.45}{0.375} = 1.2}$$

The PFTR has a removal efficiency 20% greater than the CSTR.

10.5 Self Assessment Questions

SAQ 10.1 A batch sedimentation process has to achieve a 20 $mg.l^{-1}$ discharge limit for suspended solids. The influent concentration of suspended solids is 400 $mg.l^{-1}$. The sedimentation process has a rate constant, k, of $0.5 \times 10^{-3}.s^{-1}$. What is the time required to achieve the desired discharge limit?

SAQ 10.2 If a continuous process treating wastewater has a reactor volume of 2 m^3 and a residence time of 3 h, how much wastewater would be treated per day?

SAQ 10.3 An activated sludge process with a volume V_1 has to achieve 90% degradation of the influent organics. A second activated sludge process with a volume V_2 has to achieve 99% removal. What is the ratio of V_1/V_2. (Assume 1st order reaction kinetics, the same flowrate, Q, and the same initial pollutant concentration, C_0, for both reactors.)

SAQ 10.4 A single activated sludge reactor with a 180 m^3 volume is being replaced by a train of reactors in series. The reactors are made from prefabricated 15 m^3 tanks. The influent flowrate, Q, is 0.05 m^3s^{-1} and the reaction rate constant, k, is $2.5 \times 10^{-3}s^{-1}$. The influent concentration, C_0 is 100 $mg.l^{-1}$. Approximately how many (N) reactors are needed in the new train to meet the discharge limit of 12 $mg.l^{-1}$?

SAQ 10.5 Approximately how many times bigger is the volume of the old single reactor compared to the combined volume of the replacement train of reactors in SAQ 10.4?

10.6 Nomenclature

V	reactor volume	m^3
r	volumetric rate of reaction	$kg.m^{-2}\,s^{-1}$
M	molar concentration	$mole.m^{-3}$
N	number of continuous reactors in a train	
C	concentration	$kg.m^{-3}$
C_0	initial or influent concentration	$kg.m^{-3}$
C_F	final or effluent concentration	$kg.m^{-3}$
t	time	s
k	reaction rate constant	s^{-1}
Q	flowrate	$m^3.s^{-1}$
τ	residence time	s
τ_{CSTR}	residence time in a CSTR	s
τ_{PFTR}	residence time in a PFTR	s

10.7 Solutions to Exercises

Ex 10.1 From Eqn 10.2 : $-r = k.C = 0.002 \times 25\,mg.l^{-1}.s^{-1}$
$= 0.05\,mg.l^{-1}.s^{-1}$
$= 0.05\,g.m^{-3}.s^{1}$

Ex 10.2 Answer 3. For example, in an activated sludge process with 1st order kinetics a pollutant would be removed at a rate proportional to the concentration of the pollutant remaining.

Ex 10.3 Answer 2 and 3. Batch operations are difficult to automate and thus require relatively more labour than continuous processes.
As feed is added only at the start of a batch process product concentration will increase with reaction time up to the point when all the reactant has been utilised.

Ex 10.4 $C_0 = 100$ cfu/ml; $C_F = 0.1$ cfu/ml; $k = 0.089\,min^{-1}$
Eqn 3.3: $-kt = \ln(C_F/C_0)$
$-kt = \ln 0.001 = -6.9$
$t = -6.9/-0.089 = 77$ min

Ex 10.5 Answer 1. In a batch process reactant concentration is highest at the start. When 1st order kinetics prevail the reaction rate (i.e. the rate at which product is formed) will be greatest at the start.

Ex 10.6 Answer 3. A PFTR is a continuously fed process. At any one point in the reactor the concentration of organics would be the same regardless of time. The concentration of organics is highest at the influent end of the reactor and lowest at the effluent end.

Ex 10.7 Answers 2 and 3. In a PFTR the phenolic pollutants concentration would change axially. Batch reactors are completely mixed, thus the effluent concentration, or more precisely the concentration at the termination of operation, would be the same as the reactor concentration. A CSTR is also completely mixed and has a reactor concentration equal to the effluent concentration.

Ex 10.8 Answer 1. Unless the biofilm thickness is so excessive that natural sloughing of the biofilm occurs the medium will prevent biomass being continuously wasted in the reactor outlet. Periodic cleaning of the medium is carried out, by means of backwashing with treated water, to control biofilm thickness.

Ex 10.9 99% removal means $C_0 = 100$ and $C_F = 1$.
From Eqn 3.9, for single CSTR; $\tau_{CSTR} = 99$
From Eqn 3.22 for 6 CSTRs; $\tau_{CSTR} = 6 \times (100^{(1/6)} - 1)/k$
$= 6.9/k$
From Eqn 3.15 for PFTR; $\tau_{PFTR} = -\ln 0.01/k = 4.6/k$
So volumes are in the ratio 99 : 6.9 : 4.6, that is 21.5 : 1.5 : 1.

11. Engineering Hydraulics

Contents

Aims and Objectives

This unit is an introduction to Engineering Hydraulics which you will need to understand in order to complete future Modules, assignments and projects.

After studying these notes you should be able to:

1. explain in your own words the following terms and concepts:

   Pressure and Head
   Head Loss
   Pipe Friction
   Pumping
   Porous Media

2. carry out calculations for predicting:

   Head loss through pipework systems
   Head loss in channels
   Head loss over weirs
   Head loss through porous media beds
   System curves

It is not essential for you to memorise any of the equations or constants.

It is important that you are able to correctly complete all the self assessment questions at the end of these notes.

Essential Prerequisites

Prior to undertaking this module it is important that you have completed and understood the following Unit:

Unit 7 Introduction to Process Engineering

11.1 Introduction

All water treatment plants involve the movement of fluids. This may be the bulk movement of water, the precise control of small volumes of chemical solutions or the introduction of air into a process stream. When water moves through pipes, channels or unit processes the movement is accompanied by a loss of energy as a result of friction between the water and the pipe or media in the unit process.

Water naturally flows from high level to low level under gravity and, where possible, engineers take advantage of this in the design of works but sometimes it is necessary to provide additional energy by pumping. The use of gravity flow and the specification of pumps requires calculation of pressure losses through the pipework and unit operations and the preparation of an hydraulic profile of the plant. This Unit provides a brief introduction to fluid mechanics and to the hydraulic calculations involved in water and wastewater treatment plant.

Section 11.2 requires a study time of about 1 hour

11.2 Fluid Mechanics

11.2.1 Pressure

At any depth or height, h m, within a fluid it exerts a force normal to a solid boundary or on any non-vertical plane through the fluid. The magnitude of the force is a function of the mass of fluid above the point of measurement. For an area A m^2 at depth d, the force is given by:

$$\mathbf{F = Ah\rho g} \tag{11.1}$$

where A = area in m^2
h = fluid depth in m
ρ = fluid density in $kg.m^{-3}$
g = gravitational constant, 9.81 $m.s^{-2}$

Dividing the force by the area over which it acts gives the *hydrostatic pressure*, P_h, in Nm^{-2}:

$$\mathbf{P_h = h\rho g} \tag{11.2}$$

The unit $N.m^{-2}$ is too small to be of practical use and pressures are conventionally measured in derived units such as bar or kPa, $kg.cm^{-2}$. Some of the most commonly used pressure units are tabulated in Table 11.1.

Table 11.1 – Pressure units equivalent to 1 bar

atmosphere	0.987 atm
Bar	1.000 bar
kilogram per square centimetre	1.020 $kg.cm^{-2}$
kilogram per square metre	10200 $kg.m^{-2}$
feet of water head (at 20°C)	33.460 ft
metres of water head (at 20°C)	10.197 m
millimetres of mercury	750 mmHg
newton per square metre	100,000.000 $N.m^{-2}$
kiloNewton per square metre	100.000 $kN.m^{-2}$
pascal	100,000.000 Pa
kiloPascal	100.000 kPa
pounds per square inch	14.504 psi or lb/in^2

PRESSURE MEASUREMENT
It is essential to be aware of the difference between *gauge* and *absolute* pressure measurement.
$P_{abs} = P_h + P_{atm}$

If there is a gaseous atmosphere above the free fluid surface then the *absolute* pressure, P_{abs}, at depth d will be the hydrostatic pressure plus the pressure of that atmosphere, P_{atm}:

$$\mathbf{P_{abs} = P_{atm} + P_h} \tag{11.3}$$

In the case where P_{atm} is normal atmospheric pressure (about 1bar) then the value of P_h is called the *gauge pressure*, because it is the pressure which would be measured by a pressure gauge calibrated to read zero in the normal atmosphere. It is most important, when dealing with pressure calculations, to know whether the pressure is a gauge pressure or absolute pressure. Note that absolute pressures are always positive but gauge pressures may be zero or negative. It is usual to indicate whether a pressure is gauge or absolute by suffixing a or g to the unit as in *barg* and *bara*. However, the g is often omitted and, unless otherwise indicated, it is safe to assume that *bar* means *barg*.

Ex 11.1
Calculate the pressure in barg at a depth of 25 m in sea water which has a Specific Gravity (density of liquid/density of water) of 1.02.

It is often convenient in water engineering calculations to use the depth of water as a measure of pressure, or *head* of water.

From Equation (11.2) the head, h, is:

$$h = P_h/(\rho g) \qquad (11.4)$$

Water has a density of 1000 kg/m^3 so 1 barg is approximately 10 m of water. This will not be the case when dealing with other liquids, particularly concentrated chemicals which have significantly higher densities. For example 98% sulphuric acid has a density of 1840 kg/m^3 so 1barg is about 5.4 m of sulphuric acid.

11.2.2 Viscosity

Viscosity is a measure of how fluids behave when they are moved and arises from their molecular structure. Easily flowing fluids like water or gasoline have low viscosity, whilst treacle and glues have high viscosities. In fluid mechanics terms, the viscosity relates the *rate of deformation* or *velocity gradient* of the fluid, dv/dh, to an applied *shear stress*, τ, by *Newton's Law*:

$$\tau = \mu \, (dv/dh) \qquad (11.5)$$

where μ = absolute viscosity in N.s.m^{-2} or kg.m^{-1}.s^{-1} or poise (10 P = 1 kg.m^{-1}.s^{-1}). The velocity gradient dv/dh is the change in flow velocity, v, with liquid depth, h. Viscosity falls with increasing temperature but at 20°C the value for water is 1.0×10^{-3} kg.m^{-1}.s^{-1} and for air 1.8×10^{-5} kg.m^{-1}.s^{-1}.

VISCOSITY
The viscosity of water varies with temperature by about 2% per degree.
Hence, the viscosity at T°C can be approximated by:

$$\mu = 0.001 \times (1.02)^{(20-T)}$$

Fluids like water and air, which obey Newton's Law, are referred to as *Newtonian* but many fluids are *non-Newtonian*. For some fluids, non-drip paints, ketchup and waterworks sludges for example, viscosity decreases with increasing shear force (stirring), and these are called *thixotropic*. If the viscosity increases with increasing shear, the fluid is *dilatant*, but these are less common – the most obvious example being custard.

It is sometimes convenient, in fluid mechanics calculations, to use the ratio of absolute viscosity to density and this ratio is called the *kinematic viscosity*, ν, defined by:

$$\nu = \mu/\rho \qquad (11.6)$$

Kinematic viscosity is measured in m^2.s^{-1} or Stokes (1 St = 10^{-8} m^2.s^{-1}).

11.2.3 Boundary Layer

Consider a volume flow of a fluid, Q m^3/s, flowing in a pipe of cross section area A m^2. The *free stream velocity* will be:

$$v = Q/A \qquad (11.7)$$

Ex 11.2
Calculate the free stream velocity of water flowing at 250 m^3/h in a 200 mm diameter pipe.

However, because the fluid is viscous, the fluid close to the wall of the pipe will move much more slowly than the free stream velocity and there will be a *velocity profile* across the pipe as shown in Figure 11.1. The slow moving layer of fluid close to the wall is called the *boundary layer* and its thickness is defined as that at which the velocity becomes equal to 99% of the free stream flow.

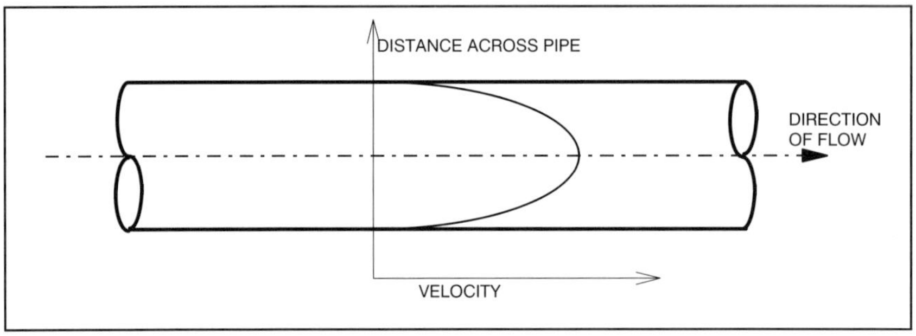

Fig. 11.1 – Velocity profile in laminar flow

REYNOLDS NUMBER

$Re = \rho dv/\mu$

Reynolds Number predicts whether flow is turbulent or laminar and is an important *dimensionless group* in many hydraulic calculations.

If the free stream velocity is increased this orderly *laminar flow* pattern begins to break up and the flow becomes *turbulent*. Reynolds' work in 1883 established that a dimensionless quantity, the *Reynolds Number* Re, could be used to predict the type of flow.

$$\mathbf{Re = \rho dv/\mu} \qquad \textbf{(11.8)}$$

where ρ = density
μ = absolute viscosity
d = a characteristic dimension (eg the diameter of a pipe)
v = free stream velocity

Ex 11.3
Calculate the Reynolds Number for the flow in Ex 2. Is the flow laminar or turbulent?

For $Re < 2000$ the flow is regarded as laminar
$Re > 4000$ the flow is generally turbulent
$2000 < Re < 4000$ flow is *transitional*

As will become evident, the Reynolds number is an important parameter in fluid mechanics calculations.

11.2.4 Bernoulli's Equation

BERNOULLI'S EQUATION
The Bernoulli equation

$H + P/\rho g + v^2/2g$

expresses the conservation of energy in fluid systems.

Consider an elevated tank full of water with a pipe connected to the bottom discharging at ground level. The water has potential energy as a result of its height. This potential energy is measured as the height of the water above ground level and is called "static head". The stored energy is increased if the tank is, for example, sealed and pressurised with air, the additional energy being calculated as the "pressure head" and is the increase in pressure above atmospheric expressed in m of water. When the water is allowed to flow out of the tank along a pipe the stored potential energy is converted into kinetic energy, imparting velocity to the water. This is measured as the "velocity head", $v^2/2g$ where v is the velocity in m/s and g is the gravitational constant 9.81 m/s^2.

The sum of these three types of energy is constant, a fact expressed in *Bernoulli's equation*:

$$\mathbf{H + P/(\rho g) + v^2/(2g) = constant} \qquad \textbf{(11.9)}$$

where H = static head, m
$P/\rho g$ = pressure head, m
$v^2/2g$ = velocity head, m

11.3 Practical Applications

Section 11.3 requires a study time of about 2 hours

11.3.1 Flow From a Tank

Consider a case where water in a sealed tank at a height of 7 m above ground level pressurised with compressed air at 3 barg (30 m) is then discharged at a height of 2 m above ground through a 100 mm diameter pipe (Figure 11.2).

Ex 11.4
What would be the velocity if the tank in Figure 11.2 were to be raised a further 5 m and opened to the atmosphere?

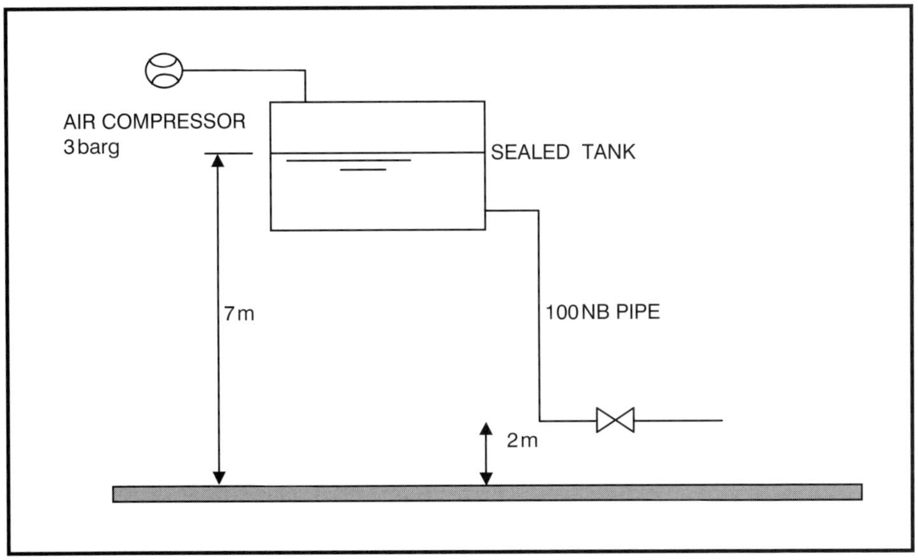

Fig. 11.2 – Flow from a tank

Initially the water is not moving so its velocity head is zero, its static head is 7 m and its pressure head 30 m. At the end of the pipe the pressure will be atmospheric and the static head will be 2 m. An energy balance shows the velocity head will be given by:

$$7 + 30 + 0 = 2 + 0 + v^2/(2 \times 9.81)$$

That is v = 26 m/s so the flow in the 100 mm pipe, which has a cross sectional area of 0.0078 m^2 would be 0.2 m^3/s or 735 m^3/h.

In fact life is not as simple as this, as will be seen later, but for the moment note that it is vital to measure all static heads from the same datum point otherwise the calculation becomes meaningless. The standard datum point for Ordnance Survey map levels is the mean sea level at Newlyn and, when this is used as a reference, levels are designates as mAOD (metres Above Ordnance Datum). In practice any fixed level can be used as a reference, for example, the level of the floor in the plant control room.

DATUM LEVEL
It is essential to ensure that all level measurements used in hydraulic calculations are based on the same datum.

Suppose that the water supply to the tank in the above example came from a river at a level of 2 m below the datum level. In order to raise the water to the tank energy for pumping the water is demanded. In this case the pump would need to generate enough static head to lift the water from −2 m to +7 m, that is 9 m. The energy input will depend on the flow required. Note that if the pump is oversized, then more energy can be imparted to the water but, because the water is being pumped into an open tank, all that will happen is that the water will splash more when it enters. The extra energy will be lost, or wasted. For this reason "breaking head" in a tank should be avoided where possible.

11.3.2 Friction Losses in Pipes

When water flows through a pipe there is, as noted earlier, friction between the wall of the pipe and the water. This slows the water down and uses up some of its energy, resulting in a loss of head along the pipe. The head loss in a pipe is related to the pipe diameter and the free stream velocity of the water in the pipe:

$$\Delta H = f\frac{l}{d}\frac{v^2}{2g} \qquad (11.10)$$

where ΔH = head loss
d = pipe diameter
l = pipe length
f = friction factor

Equation 11.10 shows that the loss of head is proportional to the flow squared so that doubling the flow increases the friction losses by four times. It is usual practice to define the head loss per unit length of pipe as the *hydaulic gradient*, s, so that Equation 11.10 becomes:

$$s = \frac{f}{d}\frac{v^2}{2g} \qquad (11.11)$$

PIPE FRICTION
Friction losses in pipework represent the energy required to make water flow along the pipe so calculation of head losses is important in sizing pumps.

The friction factor, f, in Equation 11.10 is a function of the Reynolds number and the pipe roughness, r which is a measure of the nature of the surface of the pipe. A new smooth pipe will, obviously, give a lower frictional loss than will an old rusty one. Colebrook and White carried out extensive experiments to determine head losses in commercially manufactured pipes and obtained values for the relative roughness of pipes and produced the following equation:

$$V = -2\sqrt{(2gds)}.\log\left[\frac{r}{3.7d} + \frac{2.51\nu}{d\sqrt{(2gds)}}\right] \qquad (11.12)$$

where s = hydraulic gradient $m.m^{-1}$
r = pipe roughness, m
ν = kinematic viscosity of the fluid $m^2.s^{-1}$

Typical pipe roughness values are given in Table 11.2. Note that roughness values in the literature are often given in mm.

Table 11.2 – Pipe roughness values

Pipe Type	r (m)
New plastic and non-ferrous	0.03×10^{-3}
Spun bitumen or cement lined ductile iron	0.05×10^{-3}
Steel (uncoated)	0.05×10^{-3}
Good ductile or cast iron	0.05×10^{-3}
Galvanised steel	0.15×10^{-3}
Precast concrete	0.15×10^{-3}
Tuberculated water mains up to 20 years old	$1.5 - 15 \times 10^{-3}$
Tuberculated water mains up to 50 years old	$0.3 - 30 \times 10^{-3}$

The problem with the Colebrook–White equation is that it is not explicit for s and requires an iterative solution. Over the years a variety of tables, graphs (such as that developed by Moody), nomograms (such as those published by BHRA) and computer programs have been developed for calculating friction losses. The value of f can be calculated with reasonable accuracy using:

$$\mathbf{f} = \frac{\mathbf{0.25}}{\left[\log\left(\frac{\mathbf{r}}{\mathbf{3.7d}} + \frac{\mathbf{5.74}}{\mathbf{Re^{0.9}}}\right)\right]^2} \qquad \mathbf{(11.13)}$$

At high values of Re, that is in turbulent flow regimes, this reduces to:

$$\mathbf{f} = \frac{\mathbf{0.25}}{\left[\log\left(\frac{\mathbf{r}}{\mathbf{3.7d}}\right)\right]^2} \qquad \mathbf{(11.14)}$$

which gives a value which can be used in Equation 11.11.

Suppose the water flowing from the pressurised tank at a height of 7 m in the earlier example has to flow along a length of 30 m of 100 mm pipe. If the pipe is of smooth steel then, using the appropriate r value from Table 11.2 and substituting this into Equation 11.14, the friction factor f is around 0.016. At the velocity originally calculated, 26 m/s, the head loss would be about 5.6 m per m run, that is a total of $5.6 \times 30 = 168$ m. Since the available head is only $37 - 2 = 35$ m this is clearly impossible and the flow will be reduced so that the available head is converted into velocity head. This will be around 12 m/s or a flow of about 350 m^3/h.

Ex 11.5
Use Equation 11.11 to estimate the head loss along 2 km of 300 mm diameter ductile iron pipe flowing at 320 m^3/h.

11.3.3 Losses in Pipe Fittings

Restrictions in the pipe, like changes of direction, orifice plates, valves, tee pieces with flows joining or leaving the mains stream and so on all cause a loss of head. Head losses through a particular pipe fitting can be estimated using:

$$\mathbf{H = kv^2/2g} \qquad \mathbf{(11.15)}$$

where k = a coefficient specific to the fitting.

For a series of fittings in a uniformly sized pipe with constant flow, the total head loss due to pipe fittings is calculated from the sum of the k values for the fittings:

$$\mathbf{H_T = \Sigma\, kv^2/2g} \qquad \mathbf{(11.16)}$$

Typical values for k for various fittings are given in Table 11.3.

LOSSES IN PIPE FITTINGS
Energy is expended at every change in direction of flow and where there are restrictions in pipes. Adding unnecessary fittings simply means that pumping costs will be increased.

Table 11.3 – Fitting head loss coefficients

Fitting	K
Sharp edged entrance	0.50
Bellmouth entrance	0.05
Close radius 90 deg bend	0.75
Close radius 45 deg bend	0.30
Long radius 90 deg bend	0.40
Long radius 45 deg bend	0.20
Tee with flow in line	0.35
Tee with flow in branch	0.80
Reducer	0.15

Ex 11.6
What would be the additional head loss if four long radius 90° bends and two fully open butterfly valves were to be installed in the pipeline in Ex 11.5?

Table 11.3 – (continued)

Fitting	K
Sudden contraction 3:1	0.45
Sudden contraction 5:1	0.50
Sudden enlargement 1:3	0.80
Sudden enlargement 1:5	1.00
Bellmouth exit	0.20
Gate valve fully open	0.12
Butterfly valve fully open	0.45
Reflux valve	1.00

It is a good idea to use a methodical approach to pipe lines as illustrated in the Figure 11.3.

STARING POINT
END POINT
DRAWING No

Flow, Q	m^3/h	320
Pipe diameter, d	mm	300
Pipe run length, L	m	2000
Pipe roughness, r	mm	0.05
Flow velocity, v	m/s	1.26
Head loss coefficient $C = \{1/(-2\log(r/3.7d))\}^2$		0.013
System pressure at start, Hp1	m	0.00
Start level Hs1	mAOD	25.00
End level Hs2	mAOD	40.00
System pressure required at end Hp2	m	5.00

FITTING			No OFF	k		$kv^2/2g$
Sharp edged entrance			1	0.50	m	0.04
Bellmouth entrance				0.05	m	0.00
Close radius 90 deg bend				0.75	m	0.00
Close radius 45 deg bend				0.30	m	0.00
Long radius 90 deg bend			4	0.40	m	0.13
Long radius 45 deg bend				0.20	m	0.00
Tee flow in line				0.35	m	0.00
Tee flow in branch				0.80	m	0.00
Reducer				0.15	m	0.00
Sudden contraction 3:2				0.30	m	0.00
Sudden contraction 3:1				0.45	m	0.00
Sudden contraction 5:1				0.50	m	0.00
Sudden enlargement 3:4				0.20	m	0.00
Sudden enlargement 2:3				0.35	m	0.00
Sudden enlargement 1:2				0.60	m	0.00
Sudden enlargement 1:3				0.80	m	0.00
Sudden enlargement 1:5				1.00	m	0.00
Bellmouth exit			1	0.20	m	0.02
Gate valve fully open				0.12	m	0.00
Butterfly valve	100	% open	2	0.45	m	0.07
Reflux valve				1.00	m	0.00
Static mixer				5.00	m	0.00
				0.00	m	0.00
				0.00	m	0.00
				0.00	m	0.00
				0.00	m	0.00

Total headloss due to friction in fittings	$\Sigma kv^2/2g$	m	0.26
Headloss due to friction in pipework	$(1000C/d)Lv^2/2g$	m	7.82
TOTAL FRICTION LOSSES, ΔHf		m	8.08
STATIC HEAD AT END		m	0.00
PUMP HEAD NEEDED		m	28.08

					RAW WATER SUPPLY PIPE		
					HEAD LOSS CALCULATION		
0	A N Other						
REV	BY	DATE	CHK'D	DATE	SHT 1 OF 1	REF Unit E5	0

Fig. 11.3 – Typical headloss calculation sheet

If the 30 m of pipe in the example has, say, four 90° bends in it and there is a sharp-edged entrance at the tank outlet and a sudden enlargement at the end of the pipe, the total value for k would be $4 \times 0.75 + 0.5 + 1.0 = 3.5$. The head loss would be increased by a further $3.5 \times 12^2 / (2 \times 9.81) = 25$ m. Of course, in practice, the effect will be to reduce the velocity further to around 8 m/s and so the flow to about 340 m^3/h.

PIPE VELOCITIES
As a rule of thumb velocities of water in pipework will generally be in the range 0.5–2.5 m/s.

As a general rule pipes are designed for flow velocities in the range 0.5–2.5 m/s which keeps the head loss within reasonable bounds (typically in the range 0.01–1.0 m per 100 m). Large pipes are more expensive than small ones but give a lower head loss so there is a trade off of capital cost of the pipe against operating cost (pumping power) to get the water along them. Industrial plants tend to minimise capital cost whilst utilities minimise operating costs.

Higher velocities may be dictated where there are high suspended solids levels, since these need to be kept in suspension.

11.3.4 Orifice Plates

Sometimes it is necessary to introduce a head loss in order to control flow. This may be achieved by a valve or by a fixed orifice in the line. This is usually an orifice plate, that is a plate which slips between two flanges in the pipe and has a hole drilled in its centre concentric with the pipe but of smaller diameter.

ORIFICE PLATES
Orifice plates can be used to introduce additional head loss into pipework systems in order to control flow.

An orifice of cross sectional area a in a pipe of area A generates a head loss of H metres given by:

$$v = C\sqrt{\frac{2.g.H}{\rho\left(1 - \frac{a^2}{A^2}\right)}} \qquad (11.17)$$

where v = water velocity through the orifice, m/s
C = coefficient 0.61–0.62

Ex 11.7
What would be the head loss through a 15 mm diameter orifice in a 50 mm diameter pipe flowing at 8 m^3/h?

Section 11.4 requires a study time of about 1 hour

11.4 Flow in Open Channels

11.4.1 Friction Losses

In large sewage and water treatment works it is quite usual to find open concrete channels instead of pipes. Friction losses in open channels are not easy to calculate. However, they are usually low (typically less than 0.1 m/100 m) and can be found in published tables and nomograms.

Probably the simplest of the many relationships in use is the Manning–Strickler formula:

$$\mathbf{v = K\,R^{0.66} s^{0.5}} \qquad \mathbf{(11.18)}$$

where K = wall roughness constant
R = hydraulic radius, m
s = channel gradient (i.e. the slope) in $m.m^{-1}$

The hydraulic radius is the liquid cross sectional area in m^2 divided by the wetted perimeter in m. Typical values for K are given in Table 11.4.

Table 11.4 – Manning–Strickler wall roughness

Type of Finish	K
Smooth cement rendered	60
Vitrified clay	50
Concrete	40
Metal flume	30
Earth bank	20
Grassy bank	10

BAFFLES
Baffles are commonly fitted inside tanks and unit process in water works to control levels and to change the direction of flow.

Ex 11.8
Calculate the head loss through a 500 mm square penstock at a flow of 1120 m^3/h.

11.4.2 Baffles, Slots and Penstocks

Head losses are also incurred by flow over weirs and baffles, because forcing water through a constriction in the flow always requires a certain amount of pressure. The typical formulae for losses in various configurations are given below generally give sufficient accuracy for hydraulic preliminary design calculations.

A penstock is hydraulically similar to flow under a baffle or through a horizontal slot placed near the base of the tank (Figure 11.4).

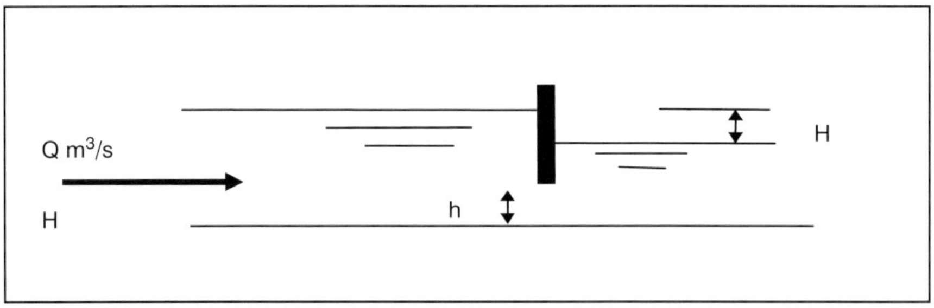

Fig. 11.4 – Flow under baffle or through horizontal slot

The head loss is given by:

$$\mathbf{Q = Chw\sqrt{(2g\ H)}} \qquad \mathbf{(11.19)}$$

where Q = flow, m^3/s
C = *discharge coefficient*, normally 0.60 to 0.62
h = height of slot or penstock, m
w = width of slot or penstock, m
H = head loss, m

For a free discharge through a penstock the value of h should be taken as the baffle height on the upstream side.

11.4.3 Vertical Baffles Normal to Flow

Where baffles are fitted normal to the flow – typically vertical baffles in a tank through which there is horizontal flow – there is a head loss as the water passes between two baffles (Figure 11.5a) or between a baffle and the tank wall (Figure 11.5b).

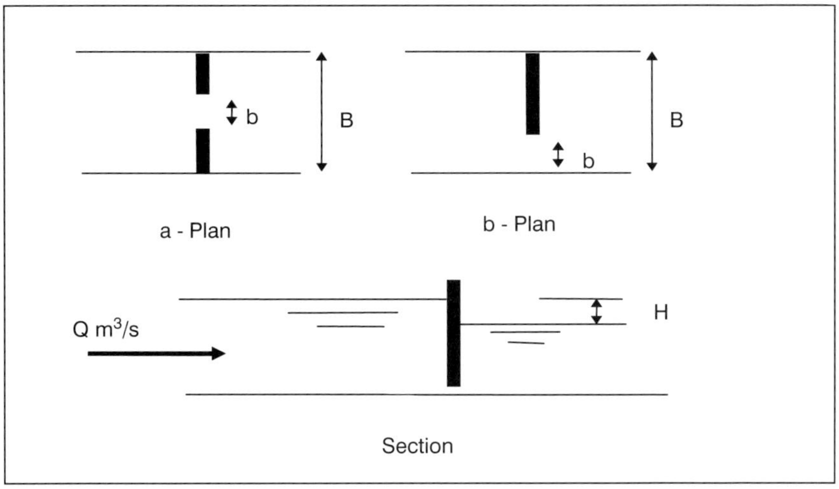

Fig. 11.5 – Vertical baffles

The calculation is a little more complicated than that shown in Section 11.4.2. If the total width of the flow channel is taken as B, and the minimum separation (baffle-to-wall or baffle-to-baffle) orthogonal to the flow – the gap width – is b, then the first step is to calculate the velocity in the gap, v m/s. A factor, M, is then given by:

$$M = 1 - (b/B)$$

Two further dimensionless factors, C_1 and C_2, are then obtained empirically from the value of M (Table 11.5). The head loss is then approximates to the term of the right-hand side of Equation 11.20.

Table 11.5 – Values for C_1 and C_2

M	C_1	C_2
0.9	0.59	0.97
0.8	0.59	0.80
0.7	0.60	0.73
0.6	0.60	0.65
0.5	0.63	0.58
0.4	0.66	0.50

$$\mathbf{H \approx C_1(v^2/2gC_2^2)} \qquad \mathbf{(11.20)}$$

11.4.4 Serpentine Baffles

Chlorine and ozone contact tanks and service reservoirs are often baffled to prevent short circuiting. The baffles may be arranged with flow over and under or around the sides (Figure 11.6). In either case the flow is serpentine and the head loss is the same.

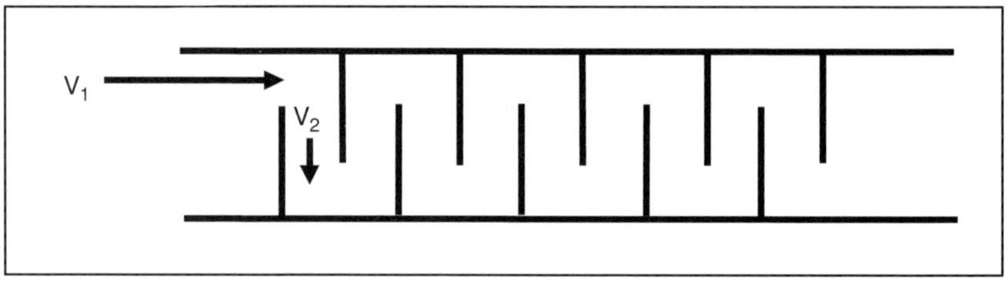

Fig. 11.6 – Serpentine baffles

If v_1 is the approach velocity (i.e. the velocity of the liquid travelling horizontally between the baffle and the wall) and v_2 the velocity between the baffles (i.e., the velocity orthogonal to the flow), then:

$$\mathbf{H \approx (1/2g) \times \{N\, v_1^2 + (N - 1)\, v_2^2\} + channel\ friction} \qquad \mathbf{(11.21)}$$

where N = number of baffles (10 in the above diagram).

11.4.5 Sharp Crested Rectangular Weirs

Weirs, forming a tank end wall over which water flows, are defined by the number of ends (N) as shown in the following front elevations:

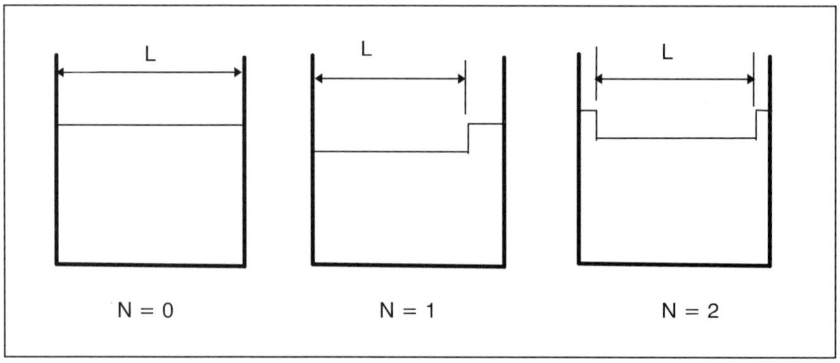

Fig. 11.7 – Sharp crested weirs, as seen from the front

Ex 11.9
Calculate the head loss when 500 m^3/h of water flows over a 1.2 m long discharge weir running the full length of the end wall of a chlorine contact tank.

The head loss for sharp crested weirs is determined from the Francis formula:

$$\mathbf{Q = K\,(L - 0.1\,N\,H)\,H^{1.5}} \qquad \mathbf{(11.22)}$$

where Q = flow, $m^3.s^{-1}$
L = effective weir length, m
K = $\frac{2}{3}C_D\sqrt{2g}$, where mean C_D (the drag coefficient) is 0.623, hence mean
K = 1.84

11.5 Flow through Porous Media

Section 11.5 requires a study time of about 1 hour

11.5.1 Granular Media

All process units will create some sort of head loss. In-line static mixers might incur around 0.5–5 m and UV lamps about the same. It is always best to seek precise head loss information from the equipment supplier if possible. Granular porous media are widely used in water and wastewater treatment and will be considered in detail. A porous medium is one which has a solid matrix containing empty spaces known as voids or pores which, for water processing, necessarily interconnect to enable fluid flow through the medium. Porous media include granular media filtration, membrane filtration, sludge dewatering, ion exchange and GAC adsorption.

Almost invariably flow through a bed of granular medium is vertical – either upflow or downflow. The medium acts as a restriction to flow and, therefore, give a head loss which is a function of the type and size of the media. Design head losses in clean units are normally in the range 1–10 m depending on whether the unit is a gravity or pressure driven. It is important to remember to allow for additional losses for fouling of filters and compaction of ion exchange resins. Typically the "clean bed head loss" might increase by 50–100% during the run.

GRANULAR MEDIA
Granular media such as sand, activated carbon, ion exchange resins and many others are commonly used in water treatment processes.

11.5.2 Darcy's Law

Darcy's Law (1856) describes laminar flow of a fluid through a porous medium:

$$v_a = k_p \frac{dH}{dl} \qquad \textbf{(11.23)}$$

where v_a = approach velocity, $m.s^{-1}$ (i.e. volumetric flow rate/gross cross sectional area)

dH/dl = hydraulic gradient in m head of water per m bed depth

k_p = permeability of the medium, $m.s^{-1}$

If the hydraulic gradient dH/dl is linear over the entire depth l of the bed, then it can be integrated to become simply h/l. This would be the case for any porous solid whose permeability is unchanged with bed depth, where permeability is the ease with which a fluid can flow through a porous medium. Permeability, which is actually the inverse of hydraulic resistance, depends upon the fluid characteristics (viscosity, density) and the characteristics of the porous material, which may change with bed depth, but is constant for a given medium and temperature.

A material may be porous but none-the-less impermeable to water, depending largely upon the pore size, the degree of hydrophobicity of the material and, most importantly, whether the pores interconnect. Pores in a bed of granular solid material, such as sand, always interconnect. The permeability, k_p, of sand used in deep bed filtration varies between 10^{-2} and $10^{-5}\,ms^{-1}$ at 10°C.

11.5.3 Kozeny–Carman Equation

The value of the permeability constant k_p in Darcy's equation is dependent on a number of physical properties of the porous/granular material. These properties principally comprise the porosity, particle shape and size range. If these properties of the porous material can be measured then it may be expected that they can be related directly to the Darcian permeability, thus allowing a direct correlation between pressure drop and flow.

The porosity ε, sometimes called “voidage”, is defined as the ratio of void or pore space to total volume occupied by the medium. In a granular material, such as sand, this value represents the space between the grains. It may be expressed as number between 0 and 1, or a percentage.

$$\varepsilon = \frac{\text{vol.voids}}{\text{vol.voids} + \text{vol.solids}} = \frac{V_v}{V_v + V} \qquad (11.24)$$

The correlation between the measurable physical properties of the porous solid and its overall permeability was first performed by Kozeny (1927) and Carman (1935). These workers considered a porous material to be made up of separate small solid particles (as for a sand filter) and the net flow to be in a single direction, even though the fluid path will necessary deviate around the solid obstructions making it *tortuous*. The individual tortuous flow paths were assumed to give the same result as a large number of parallel, identical capillary tubes with their axes in the direction of flow.

The flow through an individual cylindrical capillary was first defined by Poiseuille as being defined by the following expression:

$$\frac{H}{l} = \frac{32\mu}{\rho}\frac{V_c}{g}\frac{1}{d^2} \qquad (11.25)$$

If the internal diameter d is expressed in terms of the total volume, V, the porosity, ε, and the surface area, A, and v_c, the pore velocity, is replaced by the approach velocity v_a over the porosity ε, then Equation 11.25 becomes:

$$\frac{H}{l} = k_{k-c}\frac{\mu}{\rho}\frac{v_a}{g}\frac{(1-\varepsilon)^2}{\varepsilon^3}\left(\frac{A}{V}\right)^2 \qquad (11.26)$$

where $\mathbf{k_{k-c}}$ is the Kozeny–Carman constant and is equal to 5 for most practical cases. Experiments have shown k_{k-c} to generally fall in the range between 4 and 6.

KOZENY–CARMAN
The Kozeny–Carman equation

$$\frac{H}{l} = 5\frac{\mu}{\rho}\frac{V_a}{g}\frac{(1-\varepsilon)^2}{\varepsilon^3}\left(\frac{6}{d}\right)^2$$

predicts the head loss through a bed of granular media and provides a reasonable estimate in the absence of experimentally derived head loss data.

The equation can be further simplified for spherical grains, where:

$$\frac{A}{V} = \frac{6}{d} \qquad (11.27)$$

which gives the Kozeny–Carman equation for laminar flow in a spherical grained porous medium as:

$$\frac{H}{l} = 5\frac{\mu}{\rho}\frac{V_a}{g}\frac{(1-\varepsilon)^2}{\varepsilon^3}\left(\frac{6}{d}\right)^2 \qquad (11.28)$$

This equation can be modified to allow for grains of different shape by introducing the spehericity ϕ, though this is rarely incorporated in practice.

11.5.4 Rose Equation

Rose (1945) took a different approach and considered the drag on individual media grains and the packing arrangement of those grains, arriving at the following equation:

$$\frac{H}{l} = 0.7\frac{\mu}{\rho}\frac{v_a}{g}\frac{1}{\varepsilon^n}\left(\frac{A}{V}\right)^2 \tag{11.29}$$

which has a similar form to that of the Kozeny–Carman equation.

Ex 11.10
A typical water works sand filter uses a 600 mm deep bed of sand with effective diameter 0.6 mm and porosity 0.42. The approach velocity is 9 m/h. Use the Kozeny–Carman equation to calculate the head loss through the clean sand bed.

11.5.5 Burke–Plumber Equation

The classical expressions of Kozeny–Carman and Rose apply only to flow which is laminar. Laminar flow can assume to prevail at Reynolds numbers (Re) less than around 10, where in this instance the Re is defined as:

$$Re = \frac{dV_a\rho}{\mu(1-\varepsilon)} \tag{11.30}$$

However, the use of larger grain-size media allows greater approach velocities the Reynolds number tends to increase with increasing filter grain size, and so approach velocity. In such cases, a correction for the non-laminar flow contribution is required. The Burke–Plumber equation applies to Reynolds numbers greater than 1000:

Ex 11.11
Does the Burke–Plumber equation apply to the conditions in Ex 11.10?

$$\frac{H}{l} = \frac{1.75v_a^2(1-\varepsilon)}{dg\varepsilon^3} \tag{11.31}$$

So, the complete head-loss per unit bed depth, h/l, can be arrived at by adding the terms on the right hand side of Equations 11.28 and 11.31.

11.5.6 Practical Aspects

Porosity and grain size are important factors in determining the mean velocity and head loss through a porous medium. For a given granular medium, the voidage is approximately constant and is usually between about 0.35 and 0.6. For a bed of uniform-sized spheres the voidage is about 0.4. However, if the granular medium has a range of sizes, e.g. a graded sand, then the voidage will be reduced if the medium is well mixed because the smaller grains fill the spaces between larger ones. If the graded media becomes stratified, e.g. as a result of filter backwashing, then for each individual layer the grains are mono-sized and so will have that packing arrangement, thereby giving the voidage of a mono-sized medium for the whole bed.

Flow is affected by the actual pore size as well as the voidage. Finer grains may have the same voidage as coarser grains but the resistance to flow will be much greater do to increased surface area providing resistance to flow. Another important factor is the viscosity of the water, which is temperature dependent. This means that water flows more easily through a porous medium in summer than in winter. Lastly, the water density also changes with temperature, though not significantly. The variations are shown in Table 11.6.

Table 11.6 – Density and viscosity of water

Temperature, °C	Density, ρ, kg m^{-3}	Viscosity, μ, kg.m^{-1}.s^{-1}
10	999.73	1.3097×10^{-3}
15	999.13	1.1447×10^{-3}
20	998.23	1.0087×10^{-3}
25	997.07	0.8949×10^{-3}
30	995.68	0.8007×10^{-3}
35	994.06	0.7225×10^{-3}
40	992.25	0.6560×10^{-3}
45	990.25	0.5988×10^{-3}
50	988.07	0.5494×10^{-3}

Viscosity also increases with increasing ionic strength: a 5 wt% NaCl solution has a viscosity of 1.08×10^{-3} kg m^{-1} s^{-1} at 20°C – 8% higher than the value for pure water at the same temperature.

In practice filter beds usually stratify on backwashing, such that the particles with the larger, denser particles having the sedimentation velocity sink to the bottom of the bed and the smaller and less dense particles to the top. This gives rise to a change in the hydraulic gradient with bed depth. Moreover, all granular solids undergo compaction to varying degrees, with compaction increasing with increasing run time and increasing head loss.

Most suppliers of granular media will provide head loss data for their own media and this is normally reliable. A typical set of head loss curves for four grades of filter sand is shown in Figure 11.8.

Over the normal range of approach velocities (1–20 m.h^{-1}) and ambient temperatures, the head loss increases linearly with velocity, such that the head loss can be estimated with reasonable accurately from:

$$\mathbf{H = 1.02^{(15-T)}\, v_a\, l\, k_h} \qquad \textbf{(11.32)}$$

where H = head loss, m of water
T = temperature, °C
v_a = approach velocity, m.h^{-1}
l = bed depth, m
k_h = head loss coefficient for the medium

Typical values for k_h are given in Table 11.7.

HEAD LOSS DATA
Always use measured head loss data if it is available since real granular media are irregular and inconsistent and performance can vary widely from theory.

Table 11.7 – Head loss coefficients for media

Medium	Size, d (mm)	k_h
Ion Exchange Resin	0.4–0.5	0.12
Activated Carbon	0.6–0.7	0.03
16/30 mesh Sand	0.5–1.0	0.06
14/25 mesh Sand	0.6–1.2	0.05
8/16 mesh Sand	1.0–2.0	0.04
Grade 2 Anthracite	1.2–2.5	0.01

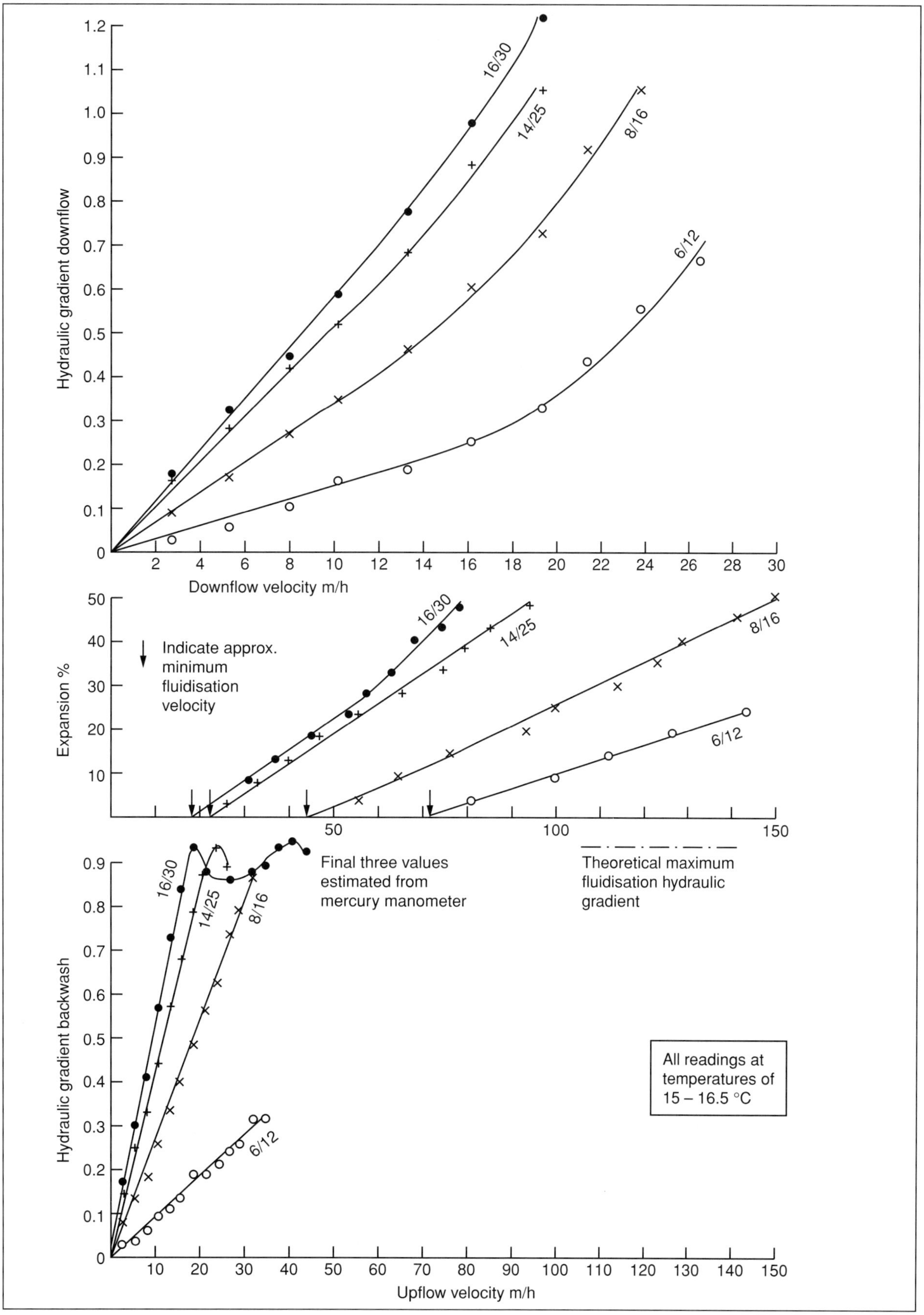

Fig. 11.8 – Laboratory test results for sand

Section 11.6 requires a study time of about 1 hour

11.6 Pumps and Pumping

11.6.1 Pump Types

With few exceptions centrifugal pumps are generally used for pumping water. Single stage pumps produce high flow and relatively low heads whilst the high heads needed for applications like reverse osmosis feed can be achieved by multistage pumps. Exceptions include raw sewage and storm water, where suspended solids may be high, flows high and heads low. Here Archimedean screw pumps are widely used. Although operating costs are relatively low, Archimedean screw pumps are large and expensive in capital cost.

Not to be confused with Archimedean screws are "screw pumps" which are more correctly called progressive cavity, hypocycloidal or Mono pumps. These positive displacement pumps are good at handling high solids and high pressures and are widely used for sludge. Where very high sludge pressures are needed, for example in sludge press feed, diaphragm or piston type positive displacement pumps are commonly used. Small diaphragm or piston pumps are also used for controlled dosing of chemicals.

11.6.2 Pump Curves

CENTRIFUGAL PUMPS
The most widely used pumps are of the centrifugal type whose performance is defined by a *performance curve.*

Centrifugal pumps impart energy into the liquid being pumped by means of a rotating *impeller.* The rotation of the impeller acts rather like a hammer thrower, creating centifugal force which throws the liquid off tangentially. The energy that is transferred from the motor via the impeller into the liquid produces a dynamic head – that is a combination of velocity and pressure heads. Since the total energy transferred remains more or less constant, the sum of the velocity and pressure heads also remains constant, that is:

$$\mathbf{P/(\rho g) + v^2/(2g) = constant} \qquad \mathbf{(11.33)}$$

which is the Bernoulli equation with H set constant. The value of v is the delivery flow divided by the area of the pump delivery connection and, since this is fixed, velocity is directly proportional to flow. As a result, if the flow delivered by the pump is restricted – for example by throttling a valve – the pressure generated will increase and vice versa. The characteristic relationship between *delivery head* and *delivery flow* is complex and varies greatly with pump design and is generally presented by the pump manufacturer in the form of a *performance curve*.

PUMP IMPELLER
The delivery flow and head from a centrifugal pump are related to impeller diameter as follows:

$$Q = k_1D$$
$$H = k_2D^2$$

The typical curve in Figure 11.9 shows the performance of a pump when fitted with five different diameter impellers. The delivery flow is proportional to the impeller diameter and the delivery head is proportional to the square of the impeller diameter. Note that there is maximum delivery pressure when the flow is zero. This is called the *closed valve head* of the pump and needs to be taken into consideration when specifying design pressures for pipes and equipment which are fed by the pump.

PUMP SPEED
The delivery flow and head from a centrifugal pump are related to pump speed as follows:

$$Q = k_1S$$
$$H = k_2S^2$$

Pump curves are produced for fixed pump speeds, usually 2900 rpm or 1450 rpm, which are the synchronous speeds of four and two-pole electric motors respectively. Frequency inverter drives allow *variable speed* operation which makes performance difficult to predict. Delivery flow is directly proportional to pump speed whilst head is proportional to the square of the speed.

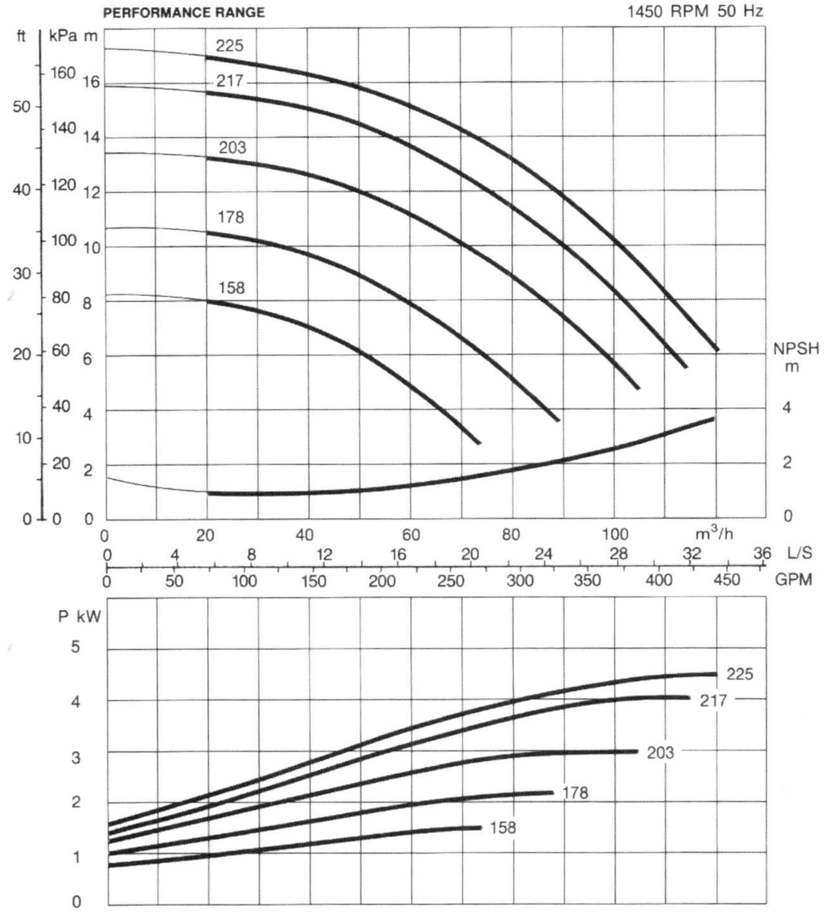

Fig. 11.9 – Centrifugal pump curve

CLOSED VALVE HEAD
The head generated by a centrifugal pump when there is no flow (ie the outlet valve is closed) is the *closed valve head* of the pump.

A *system curve* showing the pump head required for any given flow can now be constructed:

$$H_P = H + P/(\rho g) + H_f \qquad (11.34)$$

where H_P = pump head required
H_f = sum of all the head losses

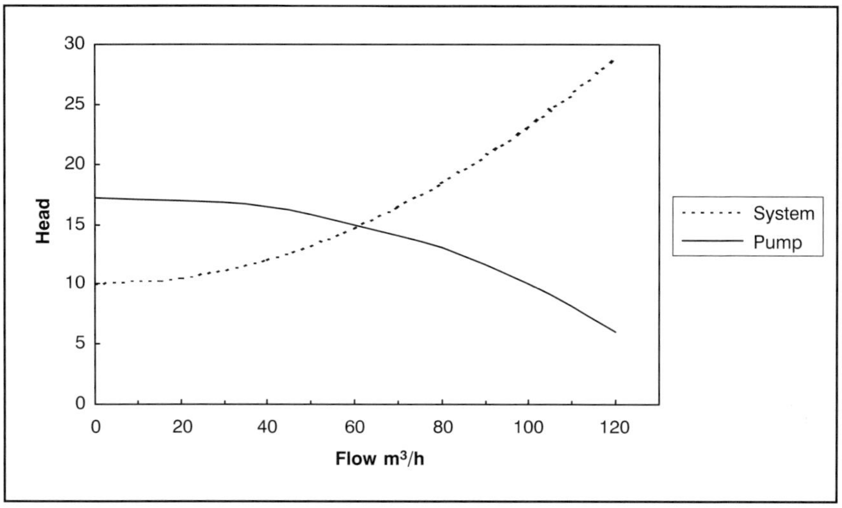

Fig. 11.10 – System and pump curves

SYSTEM CURVE
The *system curve* plots flow rate against head loss for the pipework and equipment system.

The system and pump curves intersect at the *duty point*.

The system curve, shown as the dotted line in Figure 11.10, is for a station which pumps water from a canal along a 150 mm pipe to a waterworks at 10 m above the level of the canal.

At zero flow the pump head required is the static head of 10 m. As the flow increases the friction losses increase. Since the friction losses are of the form $kv^2/2g$, the head increases with the square of the flow.

The continuous line shows the pump curve from Figure 11.10 for the 225 mm diameter impeller.

Superimposing these two curves shows that there is only a single point (the intersection) where both curves are satisfied. This is the pump *duty point* and, in this case, it is 60 m^3/h at 15 m head. So the pump selected will meet this duty point. If a flow of only 40 m^3/h was required, the system curve reveals that a pump delivery head of 12 m would be sufficient. This pump will generate too much head (about 16 m) at that flow, but some more head loss to the system could be added (by throttling a valve or adding an orifice plate) to adjust the system curve to match the pump curve. If, on the other hand, a flow of 70 m^3/h was needed the corresponding head from the system curve is 16 m. The pump will only deliver 14 m at this flow so, if we need the higher flow, we would have to change the pump for a larger one.

Increasing the head loss in the system by throttling valves or adding orifice plates means that energy is wasted and operating costs are increased, so pump selection is very important for operating cost control. Where duty points are likely to change as a result of different demands on the pump, one solution is a *variable speed drive* which allows the speed of the pump to be varied by varying the frequency of the electrical supply. The delivery flow of a pump is proportional to the speed and the delivery head is proportional to square of the speed so by adjusting the speed it is possible to achieve many different pump duties with very little difference in power consumption.

11.6.3 Pump Specification

In addition to the type of pump and the materials of construction, pump specifications should include not only the required flow and delivery head but also the suction conditions. Centrifugal pumps are not particularly good at lifting water more two or three metres. The side of the pump is at a lower pressure than the delivery and, if the pump has to lift water, for example from a sump or river, the suction gauge pressure can be negative. If the pressure at the suction is close to the vapour pressure of water then the water may start to vaporise (cavitation) causing wear on the pump impeller and inefficiency in pumping. For this reason the pump supplier will specify a requirement for the "net positive suction head" (NPSH) for the pump under a given duty.

The specification should also include the power supply specification: voltage, number of phases and frequency.

11.6.4 Net Positive Suction Head (NPSH)

NPSH

$NPSH = P_0 + H_0 - H_f - v^2/2g - P_v$

The NPSH available at a pump suction is defined as follows:

$$\mathbf{NPSH = P_0 + H_0 - H_f - v^2/2g - P_v} \qquad \textbf{(11.35)}$$

where P_0 = absolute pressure at suction, m (10.2 m for atmospheric)
H_0 = static head (negative for suction lift), m

H_f = friction loss in suction pipework, m
P_v = vapour pressure of water (see Table 11.8), m

Provided that the NPSH available is greater than the NPSH required (usually shown on the pump curve) the pump will not cavitate.

Ex 11.12
A pump draws 100 m³/h of water at 40°C from an open sump 1.5 m below the pump. The friction loss in the 200 mm diameter suction pipe is estimated at 0.3 m. What is the available NPSH?

Table 11.8 – Vapour pressure of water

Temp, °C	P_v, m	Temp, °C	P_v, m
0	0.06	40	0.74
5	0.09	50	1.23
10	0.12	60	1.99
15	0.17	70	3.11
20	0.23	80	4.73
25	0.32	90	7.00
30	0.42	100	10.00

11.6.5 Motor Power

The power absorbed when a fluid is raised through a height, h is:

$$\mathbf{E_{abs} = Q\rho gh \text{ Watts}} \qquad \textbf{(11.36)}$$

where Q = flow in $m^3.s^{-1}$
ρ = fluid density in $kg.m^{-3}$ (1000 for water)
g = gravitational constant. 9.81 $m.s^{-2}$
h = pump delivery head in m

Ex 11.13
A pump is required to deliver 120 m³/h (0.033 m³/s) against a head of 2bar (20 m). If the pump efficiency is 65% what is the minimum power required for the motor?

The efficiency, η, of a pumpset (that is the ratio of the absorbed power to the electric power input) will vary depending on the pump design and size. Typical efficiencies for centrifugal pumps range from around 30% at flows below 5 m^3/h up to 80% at flows in excess of 1000 m^3/h. The actual pump motor power is given by E_{abs}/η, that is:

$$\mathbf{E = (Q\rho gh)/(1000\ \eta) \text{ kW}} \qquad \textbf{(11.37)}$$

Section 11.7 requires a study time of about 30 minutes

HYDRAULIC PROFILE
The hydraulic profile identifies levels above datum of all free water surfaces and indicates the need for pumping.

11.7 Hydraulic Profiling

11.7.1 The Hydraulic Profile

The hydraulic profile is all about pressure and it is important that the same units of measurement are used throughout the calculation. Typically pressure might be measured in bar (more or less the same as an atmosphere), mmHg and so on but, for water systems, it is usually convenient to work in metres of water because this makes the calculations easy to visualise.

11.7.2 Methodology

Identify any fixed levels in the system. The head available at the works inlet and that required at the plant outlet are two obvious examples but the level of storage tanks may also be fixed. When identifying pump duties it is important to allow for the worst case, that is the maximum flow (which will give the maximum losses), the lowest suction level (the lowest river level for example) and the highest delivery head (the top level in a storage tank for example) *where these situations occur simultaneously*.

It is quite usual in water works design to be faced with decisions about location of structures. It may, for example, be possible to achieve gravity flow through a works by excavating to locate gravity filters at low level. As an alternative the filters could be located above ground, in a cheaper structure, and the water pumped into them. Here the capital cost of excavation has to be balanced against the running costs of the pumping system.

Identify the losses in each unit process, again it is important to allow for the highest likely flow/head loss combinations. Calculate the losses between the unit processes again considering the highest flow and remembering that head loss is proportional to flow squared.

If the head loss between two unit processes is such that it precludes gravity flow it may be worth considering reducing losses by using larger bore pipework and reducing the number of fittings.

Remember that in-line turbine or orifice plate type flow meters will impose a head loss but electromagnetic units, unless of smaller bore than the main process line, will not.

11.7.3 Presentation

The hydraulic profile of a works is best represented in the form of a process flow diagram in which levels (above a common datum) are annotated for each open process tank. The widespread use of computers for calculating hydraulic profiles has lead to the presentation in the form of a line graph and, whilst this is less easy to visualise, it is equally acceptable provided that the various items of plant are clearly identified. Examples are shown in Figures 11.11 and 11.12.

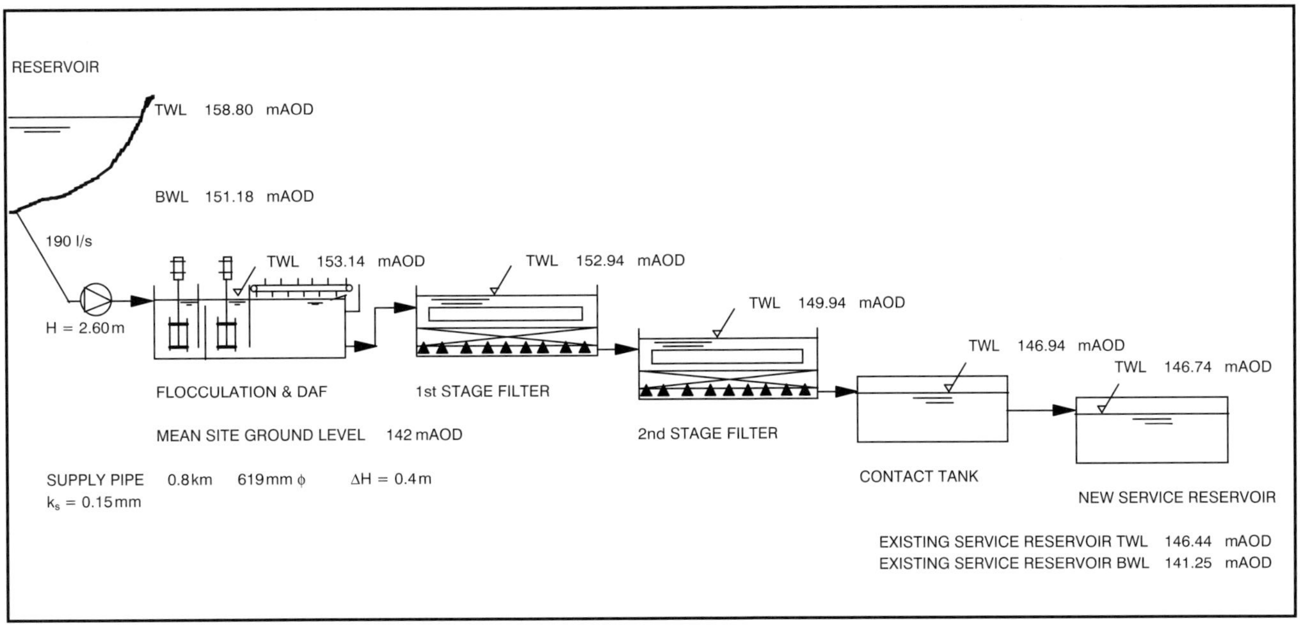

Fig. 11.11 – Typical hydraulic profile

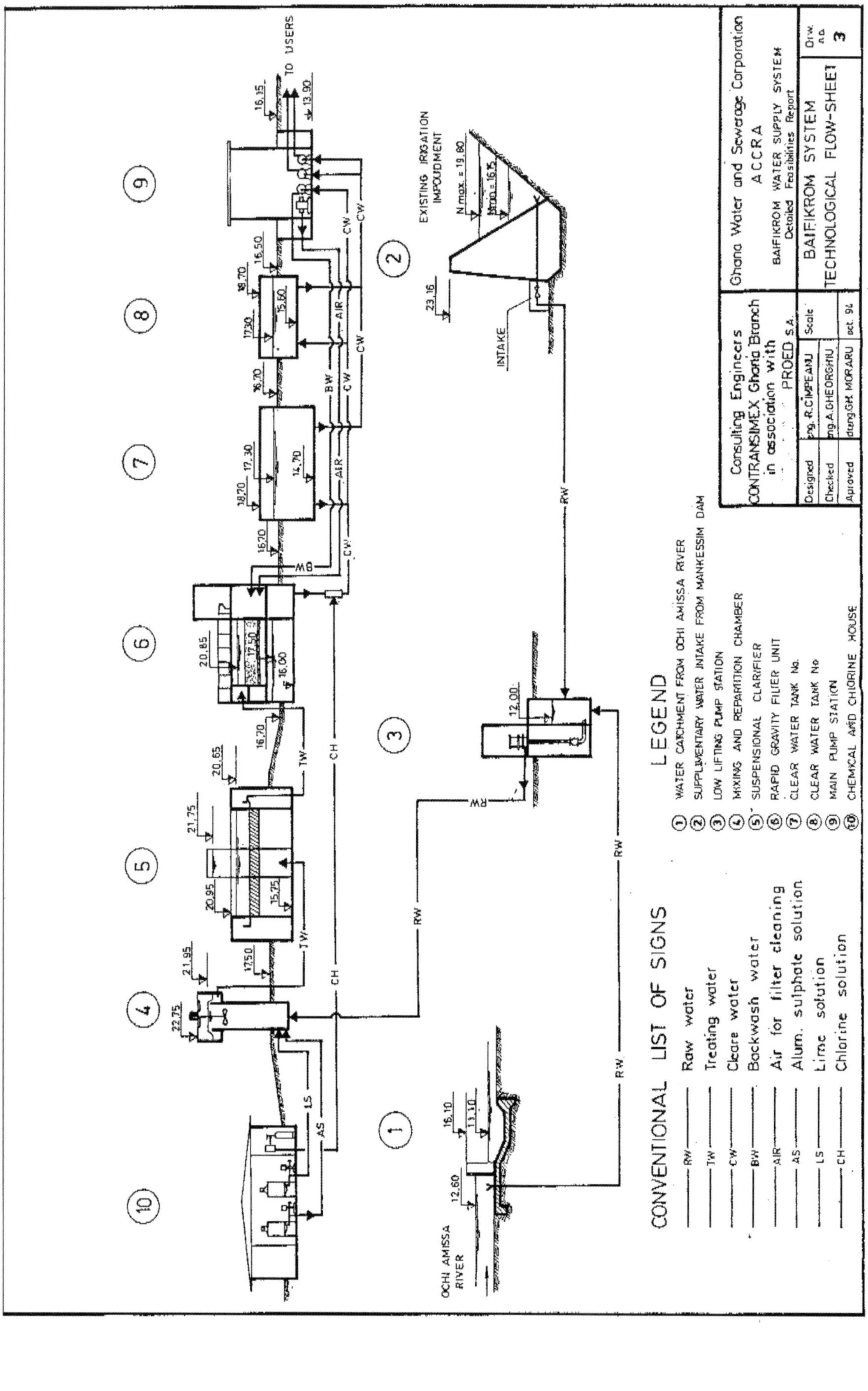

Fig. 11.12 – Typical hydraulic profile

11.8 Self Assessment Questions

SAQ 11.1 Calculate the pressure in barg at a depth of 3 m in concentrated sulphuric acid with a density of 1840 kg/m^3.

SAQ 11.2 Calculate the free stream velocity of water flowing at 25 m^3/h in a 100 mm diameter pipe.

SAQ 11.3 Calculate the Reynolds Number a flow of 25 m^3/h in a 100 mm pipe.

SAQ 11.4 What would be the velocity in a 50 mm pipe connected to an atmospheric tank elevated at 8 m above ground level if the pipe discharge point is 1m above ground level?

SAQ 11.5 Use Equation 11.11 to estimate the head loss along 5 km of 100 mm diameter ductile iron pipe flowing at 20 m^3/h.

SAQ 11.6 What would be the additional head loss if four long radius 90° bends and two fully open butterfly valves were to be installed in the pipeline in Q3.5?

SAQ 11.7 What would be the head loss through a 10 mm diameter orifice in a 50 mm diameter pipe flowing at 8 m^3/h?

SAQ 11.8 A submersible pump delivers water from a borehole at water level 70 m below ground level to a tank at 5 m above ground level. The pump is fitted with a gate valve for isolation and the 250 m long 100 mm diameter ductile iron pipe includes six long radius 90° bends. The inlet to the tank can be regarded as a sudden enlargement (1 : 5). What is the required pump delivery head at a flow of 43 m^3/h?

SAQ 11.9 Calculate the head loss through a 300 mm $\times$ 200 mm slot at a flow of 500 m^3/h.

SAQ 11.10 Calculate the head loss when 600 m^3/h of water flows over a 1.2 m long discharge weir running the full length of the end wall of a chlorine contact tank.

SAQ 11.11 What is the minimum slope required on a 500 mm wide smooth cement rendered channel to maintain a velocity of 1.5 m/s at a flow of 300 m^3/h?

SAQ 11.12 A chlorine contact tank to treat 75 m^3/h of water is 4 m long $\times$ 3 m wide $\times$ 1.0 m deep and is divided by 6 vertical baffles to form a serpentine path 21 m long $\times$ 0.5 m wide with a full width (0.5 m) weir at each end of the reactor. If the inlet weir level is 1.0 m above the tank floor what must the outlet weir level be set to so as to ensure flow through the tank?

SAQ 11.13 A typical sewage works tertiary sand filter uses a 1000 mm deep bed of sand with effective diameter 1.2 mm and porosity 0.45. The approach velocity is 7 m/h. Use the Kozeny–Carman equation to calculate the head loss through the clean sand bed.

SAQ 11.14 A pump draws 100 m^3/h of water at 80°C from an open sump 1.5 m below the pump. The friction loss in the 200 mm diameter suction pipe is estimated at 0.3 m. What is the available NPSH?

SAQ 11.15 A pump is required to deliver 2000 m^3/h (0.56 m^3/s) against a head of 5 m. If the pump efficiency is 75% what is the minimum power required for the motor?

SAQ 11.16 A borehole water at 10°C is to be treated by filtration through two pressure filters each of which treats half of the total maximum flow of 100 m^3/h. Each filter is 2.6 m in diameter and has a filter bed consisting of 300 mm depth of Grade 2 anthracite on top of a 300 mm depth of 16/30 sand. The water level in the borehole is 10 m below ground level and the filtered water is delivered to a tank 8 m above ground level. The pipework is 150 mm diameter plastic and is 300 m long with a gate valve for isolation, 8 close radius 90° bends and a tee which splits the flow to the two filters. Prepare a system curve for the plant. Use the headloss coefficients in Table 11.7 and ignore the losses in the filter inlet pipework and valves.

11.9 Nomenclature

A	cross sectional area, m^2
E	power, kW
F	force, N
H	head, m
N	number
P	pressure, $N.m^{-2}$
Q	volume flow, $m^3.s^{-1}$
R	hydraulic mean depth, m
Re	Reynolds Number
T	temperature, °C
V	volume, m^3
f	friction factor
g	gravitational acceleration, $m.s^{-2}$
h	water depth, m
k	pipe fitting head loss coefficient
k_p	permeability, $m.s^{-1}$
l	length or bed depth, m
m	mass, kg
r	pipe roughness, mm
s	hydraulic gradient, $m.m^{-1}$
v	velocity m/s
w	width, m
ρ	density, kg/m^3
μ	absolute viscosity, kg/m/s
ν	kinematic viscosity, $m^2.s^{-1}$
ε	porosity
η	fractional efficiency

11.10 Solutions to Exercises

Ex 11.1 For liquid with SG = 1.02, $\rho = 1.02 \times 1000 = 1020\,kg/m^3$
At 25 m depth, from Equation 11.2
$P = h \times \rho \times g = 25 \times 1002 \times 9.81 = 250{,}000\,Pa$
1 bar = 100,000 Pa (Table 2.1)
P = 250,000/100,000 = **2.5 bar**

Ex 11.2 $250\,m^3/h = 250/3600 = 0.069\,m^3/s$
For 200 mm (200/1000 = 0.2 m) diameter pipe the cross sectional area is: $A = \pi d^2/4 = 3.14 \times 0.2^2 \times 0.25 = 0.031\,m^2$
velocity = Q/A = 0.069/0.031 = **2.24 m/s**

Ex 11.3 $Re = \rho dv/\mu$
$\rho = 1000\,kg/m^3$
d = 0.2 m (Ex 2)
v = 2.24 m/s (Ex 2)
$\mu = 1.1 \times 10^{-3}\,kg.m^{-1}.s^{-1}$
$Re = 1000 \times 0.2 \times 2.24/(1.1 \times 10^{-3})$ = **407,000**
So flow is turbulent

Ex 11.4 Static head = 7 + 5 = 12 m
Pressure head = 0 m
Energy balance:
$12 + 0 + 0 = 2 + 0 + v^2/2g$
$v = \sqrt{(10 \times 2 \times 9.81)}$ = **14 m/s**

Ex 11.5 $s = f/d \times v^2/2g$
d = 300/1000 = 0.3 m (be careful of units)
$r = 0.05 \times 10^{-3}$ (Table 11.2)
From Eqn 11.14
$f = 0.25/[\log(r/3.7\,d)]^2 = 0.25/(-4.35)^2 = 0.013$
$Q = 320\,m^3/h = 320/3600 = 0.089\,m^3/s$
$A = \pi d^2/4 = 3.14 \times 0.3^2/4 = 0.071\,m^2$
v = Q/A = 0.089/0.071 = 1.26 m/s
$s = (0.013/0.3) \times 1.26^2/(2 \times 9.81) = 0.0035\,m/m$
pipe length = 2 km = 2000 m
total head loss = 2000×0.0035 = **7.0 m**

Ex 11.6 k values from Table:
Long radius 90° bend = 0.4, Butterfly valve = 0.45
$\Sigma k = 4 \times 0.4 + 2 \times 0.45 = 2.5$
$H = 2.5 \times 1.26^2/(2 \times 9.81)$ = **0.2 m**

Ex 11.7 $Q = 8\,m^3/h = 8/3600 = 0.0022\,m^3/s$
orifice dia = 15/1000 = 0.015 m
So, $a = \pi d^2/4 = 3.14 \times 0.015^2 \times 0.25 = 0.00018\,m^2$
v = Q/a = 0.0022/0.00018 = 12.5 m/s
$= C \sqrt{(2gH/(\rho(1 - (a^2/A^2))))}$, according to Equation 3.8.
Pipe dia = 50/1000 = 0.05 m
So, $A = \pi d^2/4 = 3.14 \times 0.05^2 \times 0.25 = 0.0019\,m^2$
Thus, $(1 - a^2/A^2) = 1 - 0.00018^2/0.0019^2 = 0.991$
Assuming C = 0.61:
$12.5 = 0.61\sqrt{(2 \times 9.81 \text{ x } H/(1000 \times 0.99))}$
So, $20.49^2 = 419.9 = (0.0198 \times H)$, thus $H = 21{,}188\,kg/m^2$
From Table 11.1, H = 21,188/10,000 = 2.12 bar
= 2.12×10.197 = **21.6 m**

Ex 11.8 $Q = 1120/3600 = 0.31\,m^3/s$
$C = 0.61$
$d = w = 500/1000 = 0.5\,m$
$0.31 = 0.61 \times 0.5 \times 0.5 \times \sqrt{}(2 \times 9.81 \times H)$
$\sqrt{}(2 \times 9.81 \times H) = 2.04$
$19.62 \times H = 4.16$
$H = 4.16/19.62 = 0.21\,m = 210\,mm$

Ex 11.9 Francis formula: $Q = K\,(L - 0.1 \times N \times H) \times H^{1.5}$
$Q = 500/3600 = 0.138\,m^3.s^{-1}$
$L = 1.2\,m$
$N = 0$ (full length weir)
$K = 1.838$
$0.138 = 1.838 \times 1.2 \times h^{0.5}$
$H^{0.5} = 0.0625$
$H = 0.004\,m = 4\,mm$

Ex 11.10 $$\frac{H}{l} = 5\frac{\mu}{\rho}\frac{v_a}{g}\frac{(1-\varepsilon)^2}{\varepsilon^3}\left(\frac{6}{d}\right)^2$$
$l = 600\,mm = 600/1000 = 0.6\,m$
$\rho = 1000\,kg/m^3$
$d = 0.6\,mm = 0.0006\,m$
$v_a = 9\,m/h = 9/3600 = 0.0025\,m/s$
$\mu = 0.001\,kg/m/s$
$\varepsilon = 0.42$
$H = 0.6 \times 5 \times (0.001/1000) \times (0.0025/9.81) \times (0.58^2/0.42^3)$
$\times (6/0.0006)^2$
$H = 0.6 \times 5 \times 1 \times 10^{-6} \times 0.00025 \times 4.54 \times 1 \times 10^8$
$H = 0.34\,m$

Ex 11.11 $Re = \rho d v_a/\mu(1 - \varepsilon)$
$\rho = 1000\,kg/m^3$
$d = 0.6\,mm = 0.0006\,m$
$v_a = 9\,m/h = 9/3600 = 0.0025$
$\mu = 0.001\,kg/m/s$
$\varepsilon = 0.42$
$Re = 1000 \times 0.0006 \times 0.0025/0.001/0.58 = 2.6$
$Re < 10$, so Burke-Plumber equation does not apply

Ex 11.12 $NPSH = P_0 + h_0 - h_f - v^2/2g - P_v$
$P_0 = 10.2\,m$ (atmospheric pressure)
$H_0 = -1.5\,m$
$H_f = 0.3\,m$
flow $= 100\,m^3/h = 100/3600 = 0.027\,m^3/s$
pipe diameter $= 200/1000 = 0.2\,m$
area $= \pi d^2/4 = 3.14 \times 0.2^2 \times 0.25 = 0.031\,m^2$
$v = 0.027/0.031 = 0.86\,m/s$
$P_v = 0.74\,m$ (Table 11.8)
$NPSH = 10.2 - 1.2 - 0.3 - 0.86^2/(2 \times 9.81) - 0.74 = 7.9\,m$

Ex 11.13 $E = (Q.\rho.g.h)/1000.\eta$ kW
$Q = 120\,m^3/h = 120/3600 = 0.033\,m^3/s$
$\rho = 1000\,kg/m^3$
$g = 9.81\,m/s^2$
$H = 2\,barg = 2 \times 10.197 = 20.2\,m$
$\eta = 65\% = 65/100 = 0.65$
$P = 0.033 \times 1000 \times 9.81 \times 20.2/(1000 \times 0.65) = 10.1\,kW$
The nearest standard sized motor would be 11 kW

12. Particle Settlement

Contents

Aims and Objectives

This Unit covers the hydraulics of particles moving in water. Separation of particles from water is used in almost every water treatment system so it is important to understand the principles.

After studying these notes you should be able to:

1. explain in your own words the following terms and concepts:

   settling velocity
   Stoke's Law

2. carry out calculations for predicting:

   settling velocity of particles

It is not essential for you to memorise any equations or constants.

It is important that you are able to correctly complete all the self assessment questions at the end of these notes.

Essential Prerequisites

Prior to undertaking this module it is important that you have completed and understood the following unit:

Unit 7 Fundamentals of Process Engineering

12.1 Introduction

Many impurities in water and wastewater are in the form of suspended particles, particularly when the water is flowing (e.g. in a river). Once quiescent or semi-quiescent conditions occur the particles will move vertically under the influence of gravity.

Usually these particles will have a density greater than that of water and will settle downwards until they form a *sludge* on the bottom. Typical examples include silt, mud and even small animals and plants, which are kept in suspension by the turbulence which results from the water's motion. Some particles are produced, either intentionally or unintentionally, by chemical conditions – typical examples are the precipitation of metal hydroxides and the use of coagulants – often in the form of *flocs* which are fluffy particles with density close to that of water. Some particles, like oil and gas bubbles, have a density less than water and they will rise to the surface and float, often forming a visible layer. In either case the rate at which the particles separate is dependent upon their size and density as well as the characteristics of the water itself.

In a later Module processes which separate particles will be considered. These include processes for separating particles whose density may be less than as well as greater than water. Sedimentation, the process for removing particles which settle in water, is the oldest and most widely used process for removing particles from liquids. It was certainly known to the ancient Egyptians and is described by Roman engineers like Pollio Vitruvius and Sextus Julius Frontius.

Section 12.2 requires a study time of about 1 hour

12.2 Sedimentation Theory

12.2.1 A General Expression

The classical theory of settling or sedimentation assumes discrete particles, such that the settling particles are not affected by either each other or the walls of the vessel.

Consider a spherical particle of mass m and density ρ_s, falling through static water. It will be subject to a downwards force, F_g, as a result of the gravitational force and this will be opposed by a buoyant force, F_b, due to displacement of the water, of density ρ, and by a drag force, F_d, which is a function of A, the particle cross-sectional area perpendicular to flow, and C_D is the drag coefficient dependent on the flow regime around the particle. The system is shown in Figure 12.1.

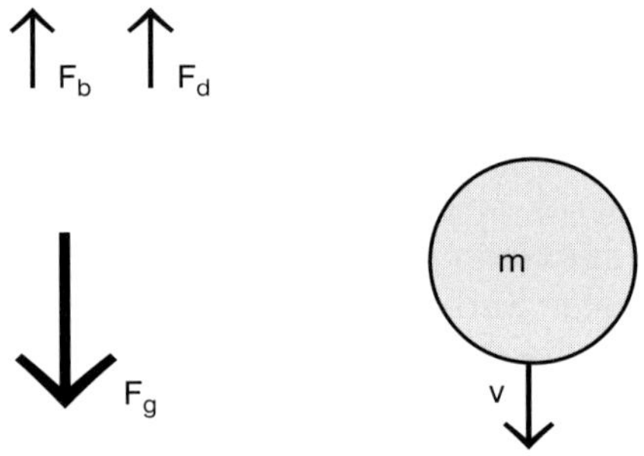

Fig. 12.1 – Forces on a particle falling through water

These forces are expressed mathematically as follows:

$$F_g = mg \tag{12.1}$$

$$F_b = \frac{mg\rho}{\rho_s} \tag{12.2}$$

$$F_d = \frac{C_D A \rho v^2}{2} \tag{12.3}$$

The resultant force on the particle is given by:

$$F = F_g + F_b + F_d = mg\left(1 + \frac{\rho}{\rho_s}\right) + \frac{C_D A \rho v^2}{2} \tag{12.4}$$

The resultant force is also defined as the rate of change of momentum of the particle, that is:

$$F = m\frac{dv}{dt} \tag{12.5}$$

If the particle is a sphere, the cross-sectional area is given by:

$$A = \frac{\pi d^2}{4} \tag{12.6}$$

and the mass is given by the volume × density:

$$m = \frac{\pi d^3 \rho_s}{6} \tag{12.7}$$

Substituting for m and A in Equation 12.4, Equation 12.6 describing the motion of the sphere becomes:

$$\frac{dv}{dt} = \frac{g(\rho_s - \rho)}{\rho_s} - \frac{3C_D\rho v^2}{4d\rho_s} \quad (12.8)$$

Initially the particle accelerates. As its velocity increases the opposing drag force also increases until it is equal to downwards forces. At this point dv/dt = 0. This means that:

$$\frac{g(\rho_s - \rho)}{\rho} = \frac{3C_D\rho v^2}{4d\rho_s} \quad (12.9)$$

The particle then continues to fall at a constant "terminal" velocity, v_t, which is given by rearranging as follows:

$$v_t = \left[\frac{4g(\rho_s - \rho)d}{3C_D\rho}\right]^{1/2} \quad (12.10)$$

This is the *general equation for settling velocity.*

TERMINAL VELOCITY
The terminal setting velocity of a particle in a fluid is given by the general equation for setting velocity.

12.2.2 Drag Coefficient

The drag coefficient C_D depends on the state of flow, which may be *viscous, transitional* or *turbulent*. The state of flow is defined by the Reynolds Number, Re, (see Unit 11) defined by:

$$Re = \frac{\rho d v_t}{\mu} \quad (12.11)$$

REYNOLDS NUMBER
The Reynolds Number (Re) is a dimensionless number which indicates whether flow is laminar (Re $<$ 5) or turbulent (Re $>$ 10^3).

In Equation 12.11, μ is the dynamic viscosity of water in $kg.m^{-1}s^{-1}$ and ρ is the density of water in $kg.m^{-3}$. Viscosity is commonly measured in Poise or centiPoise (1cP = 0.001001 $kg.m^{-1}s^{-1}$). Sometimes kinematic viscosity, ν, is used in hydraulic equations in place of dynamic viscosity when specific gravity is used in place of density. Dynamic viscosity is measured in Stokes or in $m^2.s^{-1}$ and is defined by:

$$\nu = \frac{\mu}{\rho} \quad (12.12)$$

TEMPERATURE VARIATION
The Reynolds Number (Re) is a function of the density and viscosity of water both of which vary with temperature.

Note that both viscosity and density vary with temperature (Table 12.1).

Table 12.1 – Density and viscosity of water

Temperature (°C)	Density (ρ, $kg\,m^{-3}$)	Viscosity (μ, $kg.m^{-1}.s^{-1}$)
10	999.73	1.3097×10^{-3}
15	999.13	1.1447×10^{-3}
20	998.23	1.0087×10^{-3}
25	997.07	0.8949×10^{-3}
30	995.68	0.8007×10^{-3}
35	994.06	0.7225×10^{-3}
40	992.25	0.6560×10^{-3}
45	990.25	0.5988×10^{-3}
50	988.07	0.5494×10^{-3}

Ex 12.1
Calculate the Reynolds Number for a 1mm diameter particle falling at a terminal setting velocity of 5m/h through water at 15°C.

Density is also affected by the salinity of the water, increasing by about 1% per 10 g/l salinity. Sea water, containing about 30,000 mg/l salt typically has a density of about 1030 kg.m^{-3}. This is important in the case of oil removal from sea water which is a common process in the petrochemical industry.

Typical densities of discrete particles found in water treatment are given in Table 12.2.

Table 12.2 – Density of particles

Material	Density (kg/m^3)
Air	1.3
Solvents	700–800
Vegetable oils	750–950
Mineral oils	800–980
Iron and aluminium flocs	1002–1005
Activated sludge solids	1003–1008
Wastewater organic sludge	1000–1200
Calcium precipitates	1100–1300
Sand and grit	2400–2600

The relationship between C_D and Re is shown in Figure 12.2.

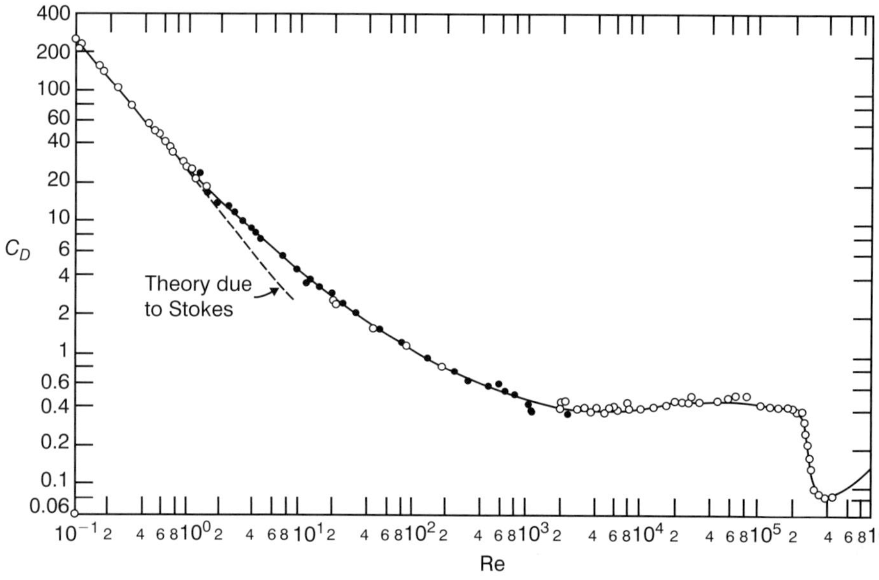

Fig. 12.2 – The dependence of C_D on Re

The turbulent and laminar flow regimes of the graph have been derived theoretically but there is no theory for transitional flow – only experimental data. Beyond Re = 10^4 the theory of ballistics applies. This region is not relevant to any water treatment applications.

12.2.3 Laminar Flow Regime

For values if Re below 2 the drag coefficient is related to the Reynolds Number by:

$$\mathbf{C_D = \frac{24}{Re} = \frac{24\mu}{\rho d v_t}} \qquad \textbf{(12.13)}$$

Substituting this into Equation 2.11 gives:

$$v_t = g\frac{(\rho_s - \rho)d^2}{18\mu} \qquad (12.14)$$

This is *Stokes' Law* and it covers most water treatment applications.

At higher Reynolds Numbers the relationship is non-linear but can be approximated by:

$$C_D = \frac{24}{Re} + \frac{3}{\sqrt{Re}} + 0.34 \qquad (12.15)$$

and, as Re becomes large, C_D approaches the constant value of 0.34 (Fig. 12.2).

Stokes' Law
Stokes' Law

$$v_t = g\frac{(\rho_s - \rho)d^2}{18\mu}$$

governs almost all gravity separation processes used in water treatment.

Ex 12.2
What is the terminal settling velocity of a 50 μm diameter silt particle of density 2600 kg/m^3 in water at 20° under Stokes' Law?

12.2.4 Turbulent Flow Regime

Under turbulent conditions, when Re is greater than 10^4, the drag coefficient is constant: $C_D = 0.4$ (Fig. 12.2). The settling velocity thus becomes:

$$v_t = \left(\frac{3.33g(\rho_s - \rho)d}{\rho}\right)^{1/2} \qquad (12.16)$$

Equation 12.16 is *Newton's Law* and is valid when $Re > 10^4$ (or sometimes 10^3).

12.2.5 Camp's Method

From the general equation for the settling velocity:

$$C_D = \frac{4g(\rho_s - \rho)d}{3\rho v_t^2} \qquad (12.17)$$

Camp manipulated this expression to yield definitions of different parameters. Dividing through by Re produces:

$$\frac{C_D}{Re} = \frac{4g(\rho_s - \rho)m}{3\rho^2 v_t^3} \qquad (12.18)$$

which is the "velocity term", and multiplying by Re^2 produces:

$$C_D Re^2 = \frac{4g(\rho_s - \rho)d^3}{3\mu^2} \qquad (12.19)$$

which is the "diameter term". Standard graphs of C_D/Re and C_DRe^2 against Re are available in the literature.

Knowledge of the velocity (e.g. by measuring settling velocity through a column of water) allows the velocity term to be calculated, and hence Re and thus the particle diameter. Conversely, knowledge of the diameter allows the diameter term to be calculated, and thus the particle velocity.

It has to be borne in mind that, when taking measurements, it takes a certain amount of time for the particle to reach its terminal velocity. In fact that time is infinite, but for practical purposes we can assume it has reached v_t when $v = 0.99v_t$.

12.3 Self Assessment Questions

SAQ 12.1 Calculate the Reynolds Number for a 0.4 mm diameter particle falling at a terminal settling velocity of 2 m/h through water at 40°C.

SAQ 12.2 What is the terminal settling velocity of a 450 μm diameter floc particle of density 1005 kg/m^3 in water at 20°C under Stokes' Law?

SAQ 12.3 What would be the required residence time in a 3 m deep tank to ensure removal of the particle in SAQ 12.2?

SAQ 12.4 What, in terms of the density difference and the particle diameter, is the ratio of the particle settling velocity in laminar and fully turbulent flow of water at around 20°C? *[NB Assume appropriate bulk parameter values for the water]*

SAQ 12.5 If solids having a density of 2.5 g/ml aggregate in water to form particles having mean diameter of 50 μm and a voidage of 80%, how many hours would it take these particles to settle in a tank 3 m deep filled with water at 20°C?

12.4 Nomenclature

A	area, m^2
C_D	drag coefficient, s^{-1}
F	force, N
L	length, m
N	number
Re	Reynolds number
W	width, m
d	diameter, m
g	acceleration due to gravity, $m.s^{-2}$
m	mass, kg
v	velocity, $m.s^{-1}$
v_t	terminal settling velocity of particle, $m.s^{-1}$
μ	dynamic viscosity, $kg.m^{-1}.s^{-1}$ ($1cP = 0.001\,kg.m^{-1}.s^{-1}$)
ρ	density, kg/m^3
ε	voidage
ν	kinematic viscosity in $m^2.s^{-1}$ ($1cSt = 10^{-6}\,m^2.s^{-1}$)
ω	angular velocity, $rad.s^{-1}$

12.5 Solutions to Exercises

Ex 12.1 $\mathbf{Re} = \frac{\rho d v_t}{\mu}$

$\rho = 999.73\,kg.m^{-3}$ (Table 12.1)
$\mu = 1.3097 \times 10^{-3}\,kg.m^{-1}.s^{-1}$ (Table 12.1)
$d = 1\,mm = 1 \times 10^{-3}\,m$
$v_t = 5\,m/h = 5/3600 = 1.38 \times 10^{-3}\,m.s^{-1}$
$Re = 999.73 \times 1 \times 10^{-3} \times 1.38 \times 10^{-3}/1.3097 \times 10^{-3}$
$= 1.05$

Ex 12.2 $\mathbf{v_t = g\frac{(\rho_s - \rho)d^2}{18\mu}}$

$d = 50\,\mu m = 5.0 \times 10^{-5}\,m$; $g = 9.81\,m.s^{-2}$; $\rho_s = 2500\,kg.m^{-3}$
$\mu = 1 \times 10^{-3}\,kg.m^{-1}.s^{-1}$; $\rho = 1000\,kg.m^{-3}$
$v_t = 9.81 \times (2500 - 1000) \times (5.0 \times 10^{-5})^2/18/1 \times 10^{-3}$
$= 2 \times 10^{-3}\,m/s = 7.35\,m/h$

Solutions to Self Assessment Questions

Unit 1 – Fundamental of Water Chemistry

SAQ 1.1 EW (Table 2.3) Ca 20, Mg 12, Na 23,
Cl 35.5, SO_4 48, NO_3 62
Convert to meq/l:
Ca = 76/20 = 3.8
Mg = 14/12 = 1.2
Na = 35/23 = 1.5
HCO_3 = 200/50 (from Equation 3.3) = 4
Cl = 38/35.5 = 1.1
NO_3 = 31/62 = 0.5
Total anions = 0.5 + 1.1 + 4 + SO_4 = 5.6
Total cations = 3.8 + 1.2 + 1.5 = 6.5
SO_4 = 0.9 meq/l = 0.9 × 48 mg/l = **43 mg/l "as such"**

SAQ 1.2 If all calcium converted to calcium sulphate:
1 mole Ca = 1 mole $CaSO_4$
Concn Ca = 200 mg/l "as $CaCO_3$" = 200/50 meq/l = 4 meq/l
Eq weight $CaSO_4$ = 20 + 48 = 68 g/eq = 68 mg/meq.
So, mass $CaSO_4$ = 4 × 68 = **272 mg**.

SAQ 1.3 *Oxidation number calculation*
Cl: From +1 in reactant to −1 in product, a change of −2
N: From −3 in reactant to +5 in product, a change of +8
Hence, for redox balance: *1 mole of NH_4^+ reacts with 4 moles of HOCl*

Charge balance
Reactants: (1 × +1) = +1
Products: (1 × −1) + (4 × −1) = −5 (nitrate and chloride respectively)
Hence for charge balance, add *6 moles of negative charge to reactants* (assumed to be hydroxide ion) **or** *6 moles positive charge to product* (as H^+).

Material balance
So far we have:
NH_4^+ 4HOCl ⇔ NO_3^- + $6H^+$ + $4Cl^-$
N and Cl both balance. The O and H count for the above equation is:
O: 1 mole deficit on product side
H: 2 mole deficit on product side
Hence the final balanced equation is:
NH_4^+ 4HOCl ⇔ NO_3^- + $6H^+$ + $4Cl^-$ + H_2O

Molar ratio HOCl : NH_4^+ = 4 : 1.
15 mg/l "as N" = 15/14 mM = 1.07 mM ammonia
1 mole HOCl = 1 mole OCl^-
So, 4 × 1.07 mM OCl^- required, 4.3 mM
1 mole OCl = 16 + 35 = 51 g/mole.
So, concn OCl^- = 4.3 mmol/l × 41 mg/mmol = **180 mg/l**.

SAQ 1.4 $AlCl_3$ + 3NaOH = $Al(OH)_3$ + 3NaCl (as SAE 5.1)
1 mole of $AlCl_3$ reacts with 3 moles of NaOH and produces 1 mole of $Al(OH)_3$.

MW (Table 2.1) Al 27, Cl 35.5, Na 23, O 16, H 1.
1 mole $AlCl_3 = 27 + (3 \times 35.5) = 133.5$ g
3 mole $NaOH = 3 \times (23 + 16 + 1) = 120$ g
1 mole $Al(OH)_3 = 27 + 3 \times (16 + 1) = 78$ g
133.5 g $AlCl_3$ reacts with 120 g NaOH produces 78 g $Al(OH)_3$
so 2 kg $AlCl_3$ reacts with $2 \times 120/133.5 = 1.8$ kg NaOH
produces $2 \times 78/133.5 = 1.2$ kg $Al(OH)_3$

SAQ 1.5 $AlCl_3 + 3H_2O = Al(OH)_3 + 3HCl$
1 mole $AlCl_3$ (133.5 g) produces 3 moles acid (3 g)
So, 10 mg/l $AlCl_3$ produces $= 3 \times 10/133.5 \times 10^{-3}$ M acid
$= 2.25 \times 10^{-4}$ M
Initial acid concentration $= 10^{-5}$ M (for pH 5)
Final acid concentration $= 2.25 \times 10^{-4} + 10^{-5} = 2.35 \times 10^{-4}$ M
Final pH $= -\log(2.35 \times 10^{-4}) =$ **3.63**

SAQ 1.6 15 g iron = 15/56 moles = 0.267 moles
From Equation 4.1, moles $= 0.267 = It/nF$
$t = 48 \times 60 \times 60 = 1.73 \times 10^5$ s F = 96493 Coulombs
For ferrous ion n = 2 (Fe^{2+}), so ;
$I = 2 \times 96493 \times 0.276 / 1.73 \times 10^5 =$ **0.31 A**.

SAQ 1.7 From Equation 4.2: n = 1
R = 8.314, F = 96500. E° is constant.
So, change in E $= (8.314 \times 298/(96500) \times \ln(10) = 0.059$ V
That is a tenfold increase in Fe^{2+}/Fe^{3+} will result in a **59 mV** reduction in actual electrode potential.

Solutions to Self Assessment Questions

Unit 2 – Chemical Kinetics and Equilibria

SAQ 2.1 From Equation 2.7, $k = -(1/t)l\ (C/C_o)$,
where $C/C_o = 1 - 0.36$ when $t = 10\,h$
So, $k = 0.0446/hr$; half life = ln 2/k = **15.5 hrs**
Resubstituting, $\ln(1 - 0.99) = -0.0446\,t$, so t = **103 hrs**

SAQ 2.2 From Equation 2.10, k = ln 2/45 = 0.0154 per min.
So, from Equation 2.8, $C/C_o = \exp -[0.0154 \times (3 \times 60)] = 0.064$
Thus % remaining $= 100 \times (1 - 0.064) = 93.6\%$

SAQ 2.3 For the equation $Na^+ + RH \Leftrightarrow RNa + H^+$ we can write:

$$K = \frac{[RNa][H^+]}{[Na^+][RH]} = 1.5$$

We know that $[RH]/[RNa] = 80/20 = 4$ and that $[Na^+] + [H^+] = 5\,mmol/l$. If the concentration of sodium in solution after treatment is *c* we have: $15 = (1 \times (5 - c))/(4c)$.
which, on rearrangement, gives x = 5/7 = **0.71 mmol/l**.

SAQ 2.4 The initial alkalinity is 280/61 = 4.59 mmol/l and the pH is 8.1.
The CO_2 is calculated from Eqn 2.41:
$8.1 - 6.38 = \log(4.59/[CO_2]) = 1.72$ $(4.59/[CO_2] = 52.5$ and $[CO_2] = 0.087\,mmol/l$.
From $HCO_3^- + HCl = CO_2 + H_2O + Cl^-$, 1 mol acid reduces alkalinity by one mol and increases CO_2 by one mol.
Adding x mol of HCl will change the pH in accordance with:
$pH - 6.38 = \log([HCO_3^- - x]/[CO_2 + x])$
To achieve the new pH of 7.0
$7.0 - 6.38 = \log([4.59 - x]/[0.09 + x])$
$\therefore (4.59 - x)/(0.09 + x) = 4.16$
$\therefore$ x = 0.82 mmol/l = **30 mg/l HCl**

SAQ 2.5 The water contains 4.4 mmol/l HCO_3^- and 2.5 mmol/l of Ca^{2+}.
Since calcium bicarbonate consists of 1 mol Ca^{2+} and 2 mol HCO_3^- the $Ca(HCO_3)_2$ concentration is 2.2 mmol/l.
From Eqn 2.46 each mol of $Ca(HCO_3)_2$ reacts with one of $Ca(OH)_2$, so the required dose is 2.2 mmol/l of $Ca(OH)_2$.
The Mol Wt of lime is 74 so the dose is $74 \times 2.2 = 162.8$ mg/l.
$50\,m^3$ will thus require $50 \times 162.8 = 8140\,g$ = **8.1 kg**.
Each mol of $Ca(HCO_3)_2$ precipitates 2 mols $CaCO_3$.
The Mol Wt of $CaCO_3$ is 100 so the weight of sludge produced is $100 \times 2.2 \times 2 = 440$ mg/l.
$50\,m^3$ will produce $50 \times 440 = 22{,}000\,g$ = **22 kg**.

SAQ 2.6 Rate equation is first order in Fe, thus:
$\ln(c/c_c) = -k[O_2]\,[OH^-]^2\,t$, where "c" refers to Fe concn.
pH = 7.8, so $[OH^-]$ init. $= 10^{-6.2}$
From Henry's Law: dissolved $O_2 = pO_2/H$
Since Mol Wt $O_2 = 32$, $[O_2] = pO_2/(H \times 32 \times 1000)$
$pO_2 = 0.2 \times 1.5 = 0.3$ atm, assuming air contains 20% oxygen,

From Table 2.2, $[O_2] = 0.3/(0.024 \times 32 \times 1000)$
$= 3.91 \times 10^{-4}$ M
required $(c/c_c) = 100/5000 = 0.02$.
So, $\ln 0.02 = -2.12 \times 10^{13} \times 3.91 \times 10^{-4} \times 10^{-6.2\times2} \times t$
Thus, t = 1200 s, **20 mins**.

SAQ 2.7

$Ca^{2+} = 2.5 \times 10^{-3}$ mol/l so pCa = 2.6
$HCO_3^- = 4.4 \times 10^{-3}$ mol/l so $pHCO_3 = 2.36$
From Fig 2.2, C at 10(C and TDS 400 mg/l = 2.5
$pH_s = 2.6 + 2.36 + 2.5 = 7.46$
$LSI = pH - pH_s = 7.5 - 7.46 =$ **+0.04** (scaling)
Adding 1 mmol/l of HCl reduces the bicarbonate by 1 mmol/l. So, HCO_3^- becomes 3.4×10^{-3} mol/l, $pHCO_3$ thus becomes 2.46, and so:
$pH_s = 2.46 + 2.36 + 2.5 = 7.32$
From Equation 2.32 CO_2 at pH 7.5 is 0.3 mmol/l.
After acid dosing this will increase to 1.3 mmol/l (since 1 mM acid generates 1 mM CO_2), and pH becomes:
$\log(3.4/1.3) + 6.38 = 6.8$.
So LSI becomes $6.8 - 7.32 =$ **−0.5** (corrosive).

Solutions to Self Assessment Questions

Unit 3 – Colloid and Surface Chemistry

SAQ 3.1 Equation 3.4

$\rho = 1100\,kg/m^3$; $d = 0.5 \times 10^{-6}\,m$; $\mu = 0.001\,kg.m^{-1}.s^{-1}$

$A/m = 6/(1100 \times 0.5 \times 10^{-6}) = 10{,}900\,m^2.kg$

If $m = 20\,mg = 20 \times 10^{-6}$

$A = 10{,}900 \times 20 \times 10^{-6} = 0.22\,m^2$

SAQ 3.2 From Equation 3.6 $D = (1.38 \times 10^{-23} \times 285)/(3 \times \pi \times 0.001 \times 0.5 \times 10^{-6})$

$= 8.3 \times 10^{-13}\,m^2.s^{-1}$

$t = 3600\,s$

From Equation 3.1 $x = \sqrt{2Dt} = \sqrt{(2 \times 8.3 \times 10^{-13} \times 3600)}$

$= 7.8 \times 10^{-5}\,m = 0.08\,mm$

SAQ 3.3 Equation 3.12: $I = 1/2\Sigma c_i z_i^2$

for Na_2SO_4: $c_{Na} = 2 \times 5 = 10\,mmol/l$; $z_{Na} = 1$

$c_{SO_4} = 1 \times 5 = 5\,mmol/l$; $z_{SO_4} = 2$

for $CaCl_2$: $c_{Ca} = 1 \times 2 = 2\,mmol/l$; $z_{Ca} = 2$

$c_{Cl} = 2 \times 2 = 4\,mmol/l$; $z_{Cl} = 1$

$\Sigma c_i z_i^2 = 10 \times 10^{-3} \times 1^2 + 5 \times 10^{-3} \times 2^2 + 2 \times 10^{-3} \times 2^2 + 4 \times 10^{-3} \times 1^2 = 0.042$

$\frac{1}{2}\Sigma c_i z_i^2 = 0.021$

SAQ 3.4 Debye length $= 1/\kappa$

From Equation 4.3 κ is proportional to $\sqrt{I}$

If I increases from 0.02 to 0.1 then $\sqrt{I}$ increases by $\sqrt{(0.1/0.02)} = 2.24$

Then κ increases by 2.24 and $1/\kappa$ decreases by $1/2.24 = 0.45$

So the Debye length is reduced by a factor of 0.45

SAQ 3.5 The Schulz-Hardy rule predicts that a higher ionic charge is more effective as a coagulant so, for a negatively charged colloid like silica, a trivalent cation like aluminium (Al^{3+}) or ferric iron (Fe^{3+}) will be more effective than a divalent cation like ferrous iron (Fe^{2+}).

Solutions to Self Assessment Questions

Unit 4 – Fundamentals of Microbiology

SAQ 4.1 He (Helium): it's an inert gas, and so cannot be assimilated by any organism.

SAQ 4.2 human liver cell; yeast

SAQ 4.3 Pathogenic micro-organisms: *Escheria coli; Giardia; Cryptosporidium*
The others are bacteria found in wastewater sludge anaerobic digesters.

Solutions to Self Assessment Questions

Unit 5 – Fundamentals of Biochemistry

SAQ 5.1

Adenine, DNA	Nucleic Acids
Starch	Carbohydrates
Enzymes	Proteins
Glycerol	Lipids

SAQ 5.2 Oxidoreductases – responsible for oxidation and reduction processes in the cell
Hydrolases – hydrolyse carbohydrates, fats, lipids into smaller molecules
Lyases – catalyses the addition or removal of substituent groups

SAQ 5.3 Catabolism = Degradation
Anabolism = Biosynthesis
Link ATP/Energy

SAQ 5.4 Oxygen (aerobic conditions) – end products CO_2, H_2O, inorganic compounds such as nitrate – end products CO_2, H_2O and organic compounds – end products ethanol

Solutions to Self Assessment Questions

Unit 6 – Microbial Kinetics

SAQ 6.1

$\ln(x_t/x_0) = \mu t$

$\ln 5 \times 10^8/1.5 \times 10^5 = \mu \times 6$

$\mu = 8.11/6 = \mathbf{1.35\,h^{-1}}$

$t_d = \ln 2/\mu_m = 0.693/1.35 = \mathbf{0.513\,h}$

SAQ 6.2

$\ln(x_t/x_0) = \mu t$ e.g. $x_0 = x_t/e^{\mu t}$

where $x_t = 3 \times 10^{18}$ cell ml^{-1}

find μ from $t_d = \ln 2/\mu_m$ (since culture is growing exponentially $\mu = \mu_m$)

$0.5 = \ln 2/\mu_m$

$\mu_m = 1.39\,h^{-1}$

$x_0 = 3 \times 10^{18}/e^{1.39\times24}$

$x_0 = 3 \times 10^{18}/2.81 \times 10^{14}$

$x_0 = 1.1 \times 10^4$ cells per ml (10,669.5 cells per ml)

SAQ 6.3

$m_{max}/V = k_La\,(c_{DO} \times -c_{DO,min})/q_0\,\tau$

$= (250 \times (6-1) \times 10^{-3})/8 \times 10^{-2} \times \tau$

$= 15.6\,kg.m^{-3}/\tau$

Production $= 28kg.m^{-3}.h^{-1} = m_{max}/(V \times \tau) = 15.6\,kg.m^{-3}/\tau$

$= 15.6/\tau$

$\therefore \tau = 15.6/28 = \mathbf{0.55\,h}$

Solutions to Self Assessment Questions

Unit 7 – Fundamentals of Process Engineering

SAQ 7.1 Rearrange equation to: $f = \Delta P.g.d/2.l.v^2$
P is force per unit area so has dimensions $M.L.T^{-2}.L^{-2} = M.L^{-1}T^{-2}$
g is an acceleration so has dimensions $L.T^{-2}$
d is a diameter so has dimensions L
l is a distance so has dimensions L
v is a velocity so has dimensions $L.T^{-1}$
so f has dimensions $M.L^{-1}.T^{-2}.L.T^{-2}.L.L^{-1}.L^{-2}.T^2 = M.L^{-2}.T^{-2}$

SAQ 7.2 Gauge pressure at bottom of tank = 8 m = 8/10.197 = 0.78 barg
Absolute pressure at bottom of tank = 1 + 0.78 = 1.78 bara
1.78 bara = 1.78 × 100 (from conversion table, p6) = 178 kPa

SAQ 7.3 Sludge volume flow = 120 $m^3.day^{-1}$ at density 1050 $kg.m^{-3}$ so
Mass flow of sludge = 120 × 1050 = 126,000 $kg.day^{-1}$
Solid content = 2% = 126,000 × 2/100
= 2520 $kg.day^{-1}$ = 2.52 $te.day^{-1}$

SAQ 7.4 600US gpm = 600/1.2 = 500 imp gpm
= 500/220 = 2.27 $m^3\ min^{-1}$
= 2.27 × 1000 = 2270 $kg.min^{-1}$ = 2270/60 = 37.88 $kg.s^{-1}$
45 ft = 45/3.28 = 13.7 m
Work done in 1 s = 37.88 × 9.81 × 13.7 = 5091 m.kgf
Rate of working = 5091 $m.kgf.s^{-1}$ = 5091 J/s
= 5091 W = 5.1 kW

SAQ 7.5 9′6″ × 12 + 6 = 114″ = 114/12 = 9.5 ft
17′9″ = 17 × 12 + 9 = 213″ = 213/12 = 17.75 ft
Cross sectional area = 9.5 × 17.75 = 168.6 ft^2
Flow = 1.5 × 168.6 = 252.9 gpm = 252.9 × 60 = 15,176 gal/h
= 15,176/220 = 69 $m^3.h^{-1}$ = 69 × 24 = 1655 $m^3.day^{-1}$
= 1655/1000 = 1.6 Mld

SAQ 7.6 1 kW = 1000 J/s; 1 hr = 3600 s; 1cubic metre = 1000 kg
So, 1 kWh = 1000 × 3600/1000 = 3600 J/kg
1 J = 1 $kg.m^2/s^2$, so 1 J/kg = 1 m^2/s^2
So, 1 kWh = 3600 m^2/s^2
Same units as pressure per unit density

Solutions to Self Assessment Questions

Unit 8 – Mass and Heat Balances

SAQ 8.1

50 l/s = 50 × 3600/1000 = 180 m^3/h
5 min = 5/60 = 0.083 h
Delivery in 5 min = 0.083 × 180 = 15 m^3
Total delivery in 1 h = 4 × 15 = 60 m^3
Average flow into tank = **60 m^3/h**

SAQ 8.2

1 kg sludge contains 1000 × 0.75 = 750 g water and (1000 − 750) = 250 g solids.
50% of the original water is removed, that is 0.5 × 750 = 375 g per kg of sludge.
Remaining sludge consists of (750 − 375) = 375 g water and the original 250 g solids. Total mass = 375 + 250 = 625 g so the solids content is 250/625 = 0.40 = **40%**

SAQ 8.3

Solution density = 1.12 g/cm^3 = 1120 kg/m^3
mass concentration = 45/1120 = 0.040 kg/kg, or 40 kg/te
Put mass flow of solution = Q te/h
Mass flow of salt = 40 kg/te × Q kg/h
Mass flow of salt = 5te/12 h = 5000/12 = 416.7 kg/h
∴ 40 × Q = 416.7
∴ Q = 10.4 te/h = 10.4 × 1000/1120 = **9.3 m^3/h**

SAQ 8.4

Mass flow of water in at 80 mg/kg = 100 m^3/h × 1000 kg/m^3 = 100,000 kg/h
Mass flow of water in at 10 mg/kg = Qkg/h
So, mass flow water in = 100,000 + Q = mass flow water out

Mass flow of sodium in at 110 mg/kg = 100,000 kg/h × 110 mg/kg = 11,000,000 mg/h
Mass flow of sodium in at 10 mg/kg = Q × 10 mg/kg
Total mass flow of sodium in = 11,000,000 + Q × 10
Mass flow of sodium out at 80 mg/kg = (100,000 + Q) × 80

∴ 11,000,000 + Q × 10 = (100,000 + Q) × 80
∴ Q = 3,000,000/70 = 42,860 kg/h = 42.8 m^3/h

SAQ 8.5

The sludge flow is 50,000 kg/day containing 10,000 mg/kg solids. Let filtrate flow be F kg/day at 50 mg/kg suspended solids and dewatered sludge S kg/day at 40% dry solids (= 400,000 mg/kg).
Mass balance on water is: 50,000 = S + F
Mass balance on solids is:
50,000 × 10,000 = S × 400,000 + F × 50
∴ F = 112,450 − S
∴ 50,000 × 5000 = (50,000 − S) × 50 + S × 400,000
∴ 50,000 × (5000 − 50) = S × (400,000 − 50)
∴ S = 618.8 kg/day
∴ F = 49,381 kg/day = 49.3 m^3/day

SAQ 8.6

The average rate of flow into the tank is 20 m^3/h
The average rate of flow from the tank is 15 m^3/h
Rate of accumulation = flow in – flow out
= 20 − 15 = 5 m^3/h
If the volume is 30 m^3 it will fill up in 30/5 = 6 hours

SAQ 8.7 Mass flow of acid solution = 350 × 1.1 = 385 te/h
Mass flow of H_2SO_4 = 385 × 15/100 = 57.75 te/h
= 5775 kg/h
1 kmol H_2SO_4 requires 1 kmol lime for neutralisation
(2 + 32 + 4 × 16) = 98 kg H_2SO_4 require 40 + 2 × (16 + 1)
= 74 kg $Ca(OH)_2$
∴ 5775 kg/h H_2SO_4 require 5775 × 74/98 = 4360 kg/h $Ca(OH)_2$
= 4360/1000 = 4.4 te/h

SAQ 8.8 In the reaction 98 kg H_2SO_4 + 74 kg $Ca(OH)_2$ produce
136 kg $CaSI_4$ + 36 kg H_2O
So 5775 kg H_2SO_4 +4360 kg $Ca(OH)_2$ produce
5775 × 136/98 = 8014 kg $CaSO_4$ + 5775 × 36/98
= 2121 kg H_2O

Mass flow of acid solution = 385 te/h
Mass flow of H_2SO_4 = 57.8 te/h
∴ Mass flow of water = 385 −57.75 = 327.3 te/h
Mass flow of $Ca(OH)_2$ = 4.4 te/h
Total mass flow in = 327.3 + 57.8 + 4.4 = 389.4 te/h
Mass flow of water produced = 2121 kg/h = 2.1 te/h
∴ Mass flow of water leaving = 327.3 + 2.1 = 329.4 te/h
Mass flow of $CaSO_4$ leaving = 8014 kg/h
Volume of water out = 329.4 × 1000/1000 = 329.4 m^3/h
∴ volumetric concentration of $CaSO_4$ = 8014 × 1000/329.4
= 24,330 g/m^3
Of which 2000g/l is soluble
∴ suspended solids = 24,330 − 2000 = 22,330 mg/l

SAQ 8.9 Assume mains water has zero COD.
Mains water mass flow in = 300,000 kg/day
mains water lost in evaporation = 0.1 × 300,000 kg/day
= 30,000 kg/day
Wastewater mass flow = 300,000 − 30,000 = 270,000 kg/day
Mass balance on COD including the 600 kg/day added to the system gives wastewater concentration ckg/kg:
300,000 × 0 + 600 = 270,000 × c
∴ c = 600/270,000 = 0.00022 kg/kg = 2.2g/kg = 2200 mg/l

SAQ 8.10 Influent wastewater flow 270,000 kg/day at COD 2200 mg/kg.
Filtrate flow = 0.8 × 270,000 = 216,000 kg/day
∴ Reject flow = 270,000 − 216,000 = 54,000 kg/day.
Mass balance on COD:
Filtrate COD = 0.7 × 2200 = 1540 mg/kg
∴ 270,000 × 2200 = 216,000 × 1540 + 54,000 × r
∴ r = (270,000 × 2200 − 216,000 × 1540)/54,000
= 4840 mg/kg

SAQ 8.11 Reject stream is 54000 kg/day COD 4840 mg/kg
Ferric chloride dose is 270 mg/kg
$FeCl_3 + 3H_2O = Fe(OH)_3 + 3HCl$
162.5 mg $FeCl_3$ = 107 mg $Fe(OH)_3$
270 mg/kg = 270 × 162.5/107 = 410 mg/kg suspended solids
0.8 × 4840 mg/kg COD = 3872 mg/kg = 3872/2
= 1936 mg/kg suspended solids

∴ total suspended solids created = 410 + 1936 = 2346 g/kg
∴ suspended solids in sludge = 2% = 20,000 mg/kg
∴ suspended solids in clarified water = 20 mg/kg
For clarified water flow C kg/day

$54{,}000 \times 2346 = C \times 20 + (54{,}000 - C) \times 20{,}000$

$\therefore C = 54{,}000 \times (20{,}000 - 2346)/(20{,}000 - 20)$

$\therefore C = 47{,}713$ kg/day

∴ Sludge quantity = 54,000 − 47,713 = 6287 kg/day

SAQ 8.12 Set reactor outlet COD = 0.3 × inlet COD. Increase recycle until reactor outlet COD = 10 mg/kg.
Recycle fraction = 300/100 = 300%

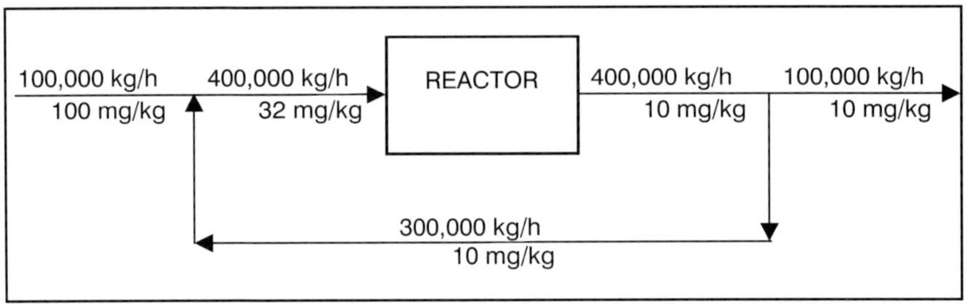

SAQ 8.13 The mass balance on COD is shown below. If a recycle in excess of 145 m^3/day is set then the COD consent of 1500 mg/l is exceeded.

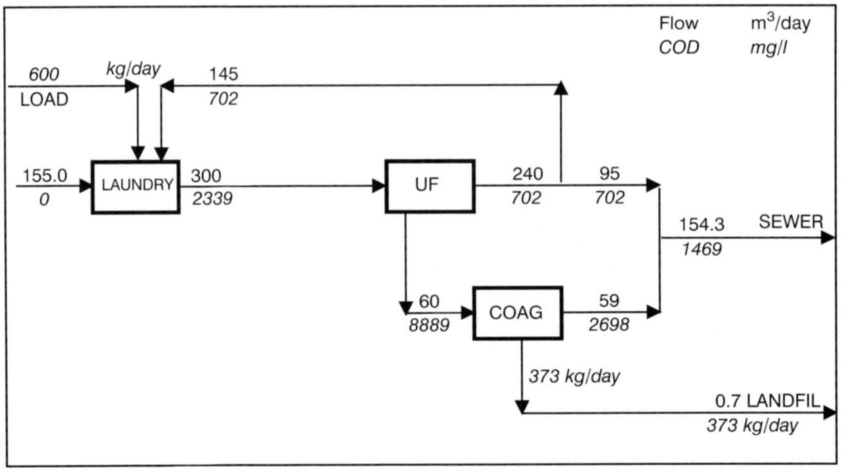

SAQ 8.14 20 g waste acid react with 1 g NaOH
∴ 1000 kg waste acid react with 1000 × 1/20 = 50 kg NaOH

$NaOH \Leftrightarrow Na^+ + OH^-$ (mol wt 40)

$Ca(OH)_2 \Leftrightarrow Ca^{2+} + 2OH^-$ (mol wt 74)

so 1 mol NaOH = 1 mol OH^- = 0.5 mol $Ca(OH)_2$

∴ 40 kg NaOH = 0.5 × 74 = 37 kg $Ca(OH)_2$

∴ 50 kg NaOH = 50 × 37/40 = 46.25 kg $Ca(OH)_2$

But the lime is only 94% $Ca(OH)_2$ so the mass of lime required to neutralise 1000 kg of acid is

46.25 × 100/94 = 49.2 kg

SAQ 8.15 Water mass flow in $= 2 \times 1000 = 2000$ kg/h
Steam mass flow in $= Q$ kg/h
Total mass flow in $= (2000 + Q)$ kg/h
$\therefore$ Total mass flow out $= (2000 + Q)$ kg/h
Steam mass flow out $= Q \times 3/100$ kg/h
Water mass flow out $= (2000 + Q) - (Q \times 3/100)$

Enthalpy flow (water) in $= 2000 \times 4.2 \times 8 = 67{,}200$ kJ/h
Enthalpy flow (steam) in $= Q \times 2707$
Total enthalpy flow in $= 67{,}200 + Q \times 2707$
$\therefore$ Total enthalpy flow out $= 67{,}200 + Q \times 2707$
Enthalpy flow out (steam) $= 2707 \times Q \times 3/100$
Enthalpy flow out (water) $= ((2000 + Q) - Q \times 3/100)) \times 4.2 \times 60$
$67{,}200 + 2707 \times Q = ((2000 + Q) - (Q \times 3/100)) \times 4.2 \times 60 + 2707 \times Q \times 3/100$
$67{,}200 + 2707 \times Q = 252 \times (2000 + 0.97 \times Q) + 81.2 \times Q$
$\therefore Q \times (2707 - 244.4 - 81.2) = 504{,}000 - 67{,}200$
$\therefore Q = 436{,}800/2381.4 = 183.4$ kg/h

SAQ 8.16 The power demand P in kJ/h for an in-line heater to heat 1500 kg/h of water from 5°C to 100°C given by:
$1500 \times 4.2 \times 5 + P = 1500 \times 4.2 \times 100$
$\therefore P = 598500$ kJ/h $= 166$ kJ/s or 166 kW. The heater is 90% efficient so power consumed $= 166 \times 100/90 = 185$ kW

SAQ 8.17 Water mass flow in $= 2 \times 1000 = 2000$ kg/h
Steam mass flow in $= Q$kg/h
Total mass flow in $= (2000 + Q)$ kg/h
$\therefore$ Total mass flow out $= (2000 + Q)$ kg/h
Steam mass flow out $= Q \times 3/100$ kg/h
Water mass flow out $= (2000 + Q) - (Q \times 3/100)$

Enthalpy flow (water) in $= 2000 \times 4.2 \times 8 = 67{,}200$ kJ/h
Enthalpy flow (steam) in $= Q \times 2707$
Total enthalpy flow in $= 67{,}200 + Q \times 2707$
$\therefore$ Total enthalpy flow out $= 67{,}200 + Q \times 2707$
Enthalpy flow out (steam) $= 2707 \times Q \times 3/100$
Enthalpy flow out (water) $= ((2000 + Q) - (Q \times 3/100)) \times 4.2 \times 60$
$67{,}200 + 2707 \times Q = ((2000 + Q) - (Q \times 3/100)) \times 4.2 \times 60 + 2707 \times Q \times 3/100$
$67{,}200 + 2707 \times Q = 252 \times (2000 + 0.97 \times Q) + 81.2 \times Q$
$\therefore Q \times (2707 - 244.4 - 81.2) = 504{,}000 - 67{,}200$
$\therefore Q = 436{,}800/2381.4 = 183.4$ kg/h

SAQ 8.18 If evaporative loss $= E$kg/h then mass balance on tower is:
$3{,}000{,}000 = Q + E$ where Q is residual circulating flow
Enthalpy balance is:
$3{,}000{,}000 \times 4.2 \times 40 = E \times 2450 + (3{,}000{,}000 - E) \times 4.2 \times 35$
$\therefore 3{,}000{,}000 \times 4.2 \times (40 - 35) = E \times (2450 - 4.2 \times 35)$
$\therefore E = 27{,}355$ kg/h $= 0.9\%$ of the circulating flow

SAQ 8.19 Total water losses $= 27{,}355 + 0.01 \times 3{,}000{,}000 = 57{,}355$ kg/h
Make-up water Cl concentration $= 50$ mg/l
Mass balance on chloride assuming concentration cmg/kg in circulating water and zero chloride in evaporating steam:
make-up $= 57{,}355 \times 50$
losses $= 27{,}355 \times 0 + 30{,}000 \times c$
$\therefore 27{,}355 \times 0 + 30{,}000 \times c = 57{,}355 \times 50$
$\therefore c = 57{,}355 \times 50/30{,}000 = 95.5$ mg/kg

Solutions to Self Assessment Questions

Unit 9 – Introductory Mass and Heat Transfer

SAQ 9.1 Using Equation 9.15

$Sh = 0.13 \times Gr^{0.3} \times Sc^{0.3}$

$Gr = l^3 g\rho\Delta\rho/\mu^2$

l = water depth = 2 m

g = gravitational constant, $9.81\ m.s^{-2}$

From Unit E1 Table 5.1 at 10°C:

$\rho = 999.7\ kg.m^{-3}$

$\mu = 1.3 \times 10^{-3}\ kg.m^{-1}.s^{-1}$

$\Delta\rho = (999.7 - 1.43) = 998.3$

$Gr = 2^3 \times 9.81 \times 999.7 \times 998.3/(1.3 \times 10^{-3})^2$

$Gr = 4.63 \times 10^{13}$

$Sc = \mu/\rho D$

From Table 9.1, $D = 2.5 \times 10^{-9}\ m^2.s^{-1}$

$Sc = 1.3 \times 10^{-3}/999.7/(2.5 \times 10^{-9})$

$Sc = 520.15$

$Sh = 0.13 \times (4.63 \times 10^{13})^{0.3} \times 520.15^{0.3}$

$Sh = 0.13 \times 12579.8 \times 6.53 = 10{,}677$

$Sh = k.l/D$

$k = Sh \times D/l$

$k = 10{,}677 \times 2.5 \times 10^{-9}/2 = 1.33 \times 10^{-5}\ m.s^{-1}$

$k_La = 1.33 \times 10^{-5} \times 3600 = \mathbf{0.05\ h^{-1}}$

SAQ 9.2 Calculate % oxygen deficit at time t D = 100 – % sat'n

Calculate D/D_0 then $\ln(D/D_0)$:

Time (min)	% sat	% D	D/D_0	$\ln(D/D_0)$
0	11	89	1	0
2	14	86	0.966292	−0.034289
4	18	82	0.921348	−0.081917
6	22	78	0.876404	−0.131928
8	25	75	0.842697	−0.171148
10	28.5	71.5	0.803371	−0.218939
12	32	68	0.764045	−0.269129
14	35.5	64.5	0.724719	−0.321971
16	38	62	0.696629	−0.361502
18	41	59	0.662921	−0.411099
20	44	56	0.629213	−0.463285
25	50	50	0.561798	−0.576613
30	55	45	0.505618	−0.681974
35	61	39	0.438202	−0.825075

Plot time, t vs $\ln(D/D_0)$:

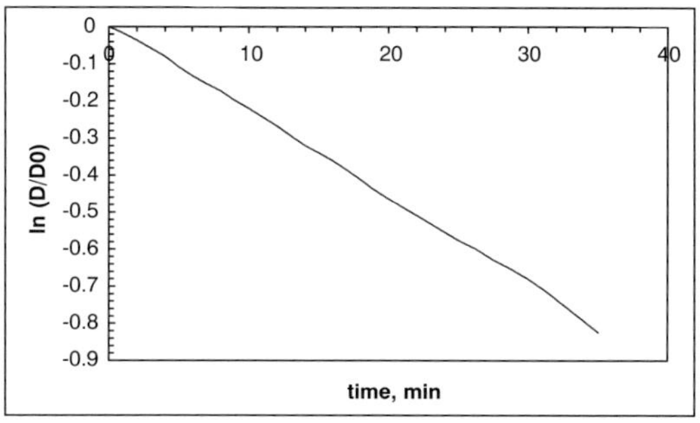

k_La = Slope = (−)1.4/h at 6°C

SAQ 9.3

Ring main is Sch 10 316L stainless steel = 2″ ID
Thickness (RGB Manual) = 0.109″
Length = 350 m
Ambient temperature = 18°C
Sanitising inlet temperature = 90°C
Heat transfer coef (IHVE Handbook) = 18 $W/m^2/°C$
Area of tube = 0.2 m^2/m
Area of run = 61.9 m^2
Temperature difference = 72.0°C
Rate of heat loss = 80.3 kJ/s
Flow rate = 13.5 m^3/h
Flow rate = 3.8 kg/s
Enthalpy of flow = 378.0 kJ/kg
Heat content of flow = flow × enthalpy = 1417.5 kJ/s
Heat content of return = 1337.2 kJ/s
Enthalpy of return = heat content/flow = 356.6 kJ/kg
Temperature of return = 84.9°C

SAQ 9.4

65 m^3/h = 18 kg/s; 30 m^3/h = 5.56 kg/s
Heat balance to find outlet temperature of hot water
$H = 18 \times 4.2 \times (30 - 10) = 5.56 \times 4.2 \times (80 - T)$
$\therefore H = 1512$ kJ/s and $T = 15$
$\Delta T_m = [(80 - 30) - (15 - 10)]/\ln(50/5) = 45/2.3 = 19.5$
$H = UA\Delta T_m \therefore A = 1512/2000/19.5 = 0.038\,m^2$
Thermal resistance of scale $= x/k_t = 0.00001/3.6$
$= 27.7 \times 10^{-6}$
U for scaled plates $= 1/[(1/2000) + 27.2 \times 10^{-6}] = 1895$
$H = UA\Delta T_m$
$\therefore \Delta T_m = 1512/1895/0.038 = 21°C$
New inlet temperature, T
$21 = [(T - 30) - (15 - 10)]/\ln((T - 30)/5)$
if $T - 30 = x$
$\ln(x/5) = (x - 5)/21$
$\therefore x = 56$
$\therefore T = 86°C$

Solutions to Self Assessment Questions

Unit 10 – Reactor Design Theory

SAQ 10.1 Rearrange Eqn 10.6 to make time, t, the subject and substitute the values for C, C_0 and k given;

$$t = \frac{\ln\left(\frac{C}{C_0}\right)}{k}$$

$$t = \frac{\ln\left(\frac{20}{400}\right)}{0.5 \times 10^{-3}} = 5991\,\text{s} \approx 100\,\text{min}$$

SAQ 10.2 Rearrange Eqn 10.8 to make flowrate, Q, the subject and substitute the values for τ and V given;

$$Q = \frac{V}{\tau} = \frac{2}{3} = 0.66\ \text{m}^3.\text{h}^{-1}$$

Multiplying 0.66 $\text{m}^3.\text{h}^{-1}$ by 24 h gives the capacity 16 m^3 per day.

SAQ 10.3 For reactors V_1 and V_2 Eqn 10.13 can be rewritten, expressing the final concentration, C_F, in terms of the initial concentration, C_0;

$$k\tau = \frac{C_0 - C_F}{C_F}$$

for V_1;

$$k\tau = \frac{C_0 - 0.1C_0}{0.1C_0}$$

for V_2;

$$k\tau = \frac{C_0 - 0.01C_0}{0.01C_0}$$

From Eqn 10.8 volume equals τ/Q. In this case Q is the same for both reactors, therefore $V = \tau$;

$$\frac{V_1}{V_2} = \frac{\tau_1}{\tau_2} = \frac{\left(\frac{C_0 - 0.1C_0}{0.1C_0}\right)}{\left(\frac{C_0 - 0.01C_0}{0.01C_0}\right)} = \frac{\left(\frac{0.9}{0.1}\right)}{\left(\frac{0.99}{0.01}\right)} = \frac{9}{99} = \frac{1}{11}$$

SAQ 10.4 To determine the number, N, of reactors required the residence time, τ, is determined using Eqn 10.8

$$\tau = \frac{\text{volume}}{\text{flow}} = \frac{V}{Q} = \frac{15}{0.05} = 300\,\text{s}$$

Equation (10.25) is then rearranged to make N the subject. This is done by first taking the log of both sides of the equation;

$$\log\left(\frac{C_0}{C_N}\right) = N\log(1 + k\tau)$$

$$N = \frac{\log\left(\frac{C_0}{C_N}\right)}{\log(1 + k\tau)}$$

substituting values for C, C_0, k and τ:

$$N = \frac{\log\left(\frac{100}{12}\right)}{\log(1 + 2.5 \times 10^{-3} \cdot 300)} = \frac{0.92}{0.24} = 3.8 \approx 4$$

SAQ 10.5 Approximately 4 reactors of 15 m^3 volume are required in the train; giving a total volume of 60 m^3. The original process had a volume of 180 m^3. Therefore, the new process is 3 times smaller.

Solutions to Self Assessment Questions

Unit 11– Engineering Hydraulics

SAQ 11.1 Liquid with density, $\rho = 1840\,kg/m^3$
At 3 m depth, from Equation 11.2
$P = h \times \rho \times g = 3 \times 1840 \times 9.81 = 55{,}150\,Pa$
1 bar = 100,000 Pa (Table 11.1)
$P = 55{,}150/100{,}000 =$ **0.54 bar**

SAQ 11.2 $25\,m^3/h = 25/3600 = 0.0069\,m^3/s$
For 100 mm (100/1000 = 0.1 m) diameter pipe the, cross sectional area is: $A = \pi d^2/4 = 3.14 \times 0.1^2 \times 0.25 = 7.8 \times 10^{-3}\,m^2$
velocity $= Q/A = 0.0069/7.8 \times 10^{-3} =$ **0.87 m/s**

SAQ 11.3 $Re = \rho dv/\mu$
$\rho = 1000\,kg/m^3$
$d = 0.1\,m$
$v = 0.87\,m/s$ (Q 3.2)
$\mu = 1.1 \times 10^{-3}\,kg.m^{-1}.s^{-1}$
$Re = 1000 \times 0.1 \times 0.87/(1.1 \times 10^{-3}) =$ **79,090**
So flow is turbulent

SAQ 11.4 Static head = 8 m
Pressure head = 0 m
Energy balance $= 12 + 0 + 0 = 1 + 0 + v^2/2g$
$v = \sqrt{(11 \times 2 \times 9.81)} =$ **14.7 m/s**

SAQ 11.5 $s = f/d \times v^2/2g$
$d = 100/1000 = 0.1\,m$ (be careful of units)
$r = 0.05 \times 10^{-3}$ (Table 11.2)
From Eqn 11.14
$f = 0.25/[\log(r/3.7d)]^2 = 0.25/(-3.86)^2 = 0.017$
$Q = 20\,m^3/h = 20/3600 = 5.5 \times 10^{-3}\,m^3/s$
$A = \pi d^2/4 = 3.14 \times 0.1^2/4 = 7.9 \times 10^{-3}\,m^2$
$v = Q/A = 5.5 \times 10^{-3}/7.9 \times 10^{-3} = 0.7\,m/s$
$s = (0.017/0.1) \times 0.7^2/(2 \times 9.81) = 0.0042\,m/m$
pipe length = 5 km = 5000 m
total head loss $= 5000 \times 0.0042 =$ **21 m**

SAQ 11.6 k values from Table:
Long radius 90° bend = 0.4, Butterfly valve = 0.45
$\Sigma k = 4 \times 0.4 + 2 \times 0.45 = 2.5$
$H = 2.5 \times 0.7^2/(2 \times 9.81) =$ **0.06 m**.

SAQ 11.7 $Q = 8\,m^3/h = 8/3600 = 0.0022\,m^3/s$
orifice dia = 10/1000 = 0.01 m
So, $a = \pi d^2/4 = 3.14 \times 0.01^2 \times 0.25 = 7.85 \times 10^{-5}\,m^2$
$v = Q/a = 0.0022/0.000079 = 28\,m/s$
$= C\sqrt{(2gH/(\rho(1 - (a^2/A^2))))}$, according to equation 11.17
Pipe dia = 50/1000 = 0.05 m
So, $A = \pi d^2/4 = 3.14 \times 0.05^2 \times 0.25 = 0.0019\,m^2$
Thus, $(1 - a^2/A^2) = 1 - (0.000079^2/0.0019^2) = 0.998$
Assuming C = 0.61:
$28 = 0.61\sqrt{(2 \times 9.81 \times H/(1000 \times 0.998))}$
So, $45.9^2 = 2107 = (0.0196 \times H)$, thus $H = 107174\,kg/m^2$
From Table 2.1, $H = 107174/10000 = 10.7\,bar = 10.7 \times 10.197$
$=$ **109 m**

SAQ 11.8 Static head at inlet = −70 m
Static head at end = +5 m
Static head = 5 − (−70) = 75 m
Flow = 43 m^3/h = 0.012 m^3/s , pipe area = 0.0078 m^2;
v = 1.53 m/s
$s = k \times v^2/2g$
r = 0.05 (Table 3.1)
$k = (1100/D) \times (1/(-2 \times \log(r/3.7/D)))^2 = 0.18$
$s = 0.18 \times 1.53^2/2/9.81 = 0.022$ m/m
L = 250 m; $h_P = 0.022 \times 250 = 5.5$ m
k (Table 11.3) long radius bend = 0.40
gate valve = 0.12
sudden enlargement = 1.00
$k_T = 6 \times 0.4 + 0.12 + 1.00 = 3.52$
$H_F = 3.52 \times 1.53^2/2/9.81 = 0.42$ m
Total pump head required = 75 + 5.5 + 0.4 = 80.9 m

SAQ 11.9 Q = 50/3600 = 0.014 m^3/s
C = 0.61
d = 200/1000 = 0.2 m
w = 300/1000 = 0.3 m
$0.014 = 0.61 \times 0.2 \times 0.3 \times \sqrt{(2 \times 9.81 \times H)}$
$\sqrt{(2 \times 9.81 \times H)} = 0.38$
$19.62 \times H = 0.14$
$H = 0.14/19.62 = 7.3 \times 10^{-3}$ m = 7.3 mm

SAQ 11.10 Francis formula: $Q = K (L - 0.1 \times N \times H) \times H^{1.5}$
Q = 600/3600 = 0.167 $m^3.s^{-1}$
L = 1.2 m
N = 0 (full length weir)
K = 1.838
$0.167 = 1.838 \times 1.2 \times H^{0.5}$
$H^{0.5} = 0.076$
H = 0.006 m = 6 mm

SAQ 11.11 Flow = 300 m^3/h = 300/3600 = 0.083 m^3/s
velocity = 1.5 m/s
∴ cross sectional area of flow = 0.083/1.5 = 0.056 m^2
channel width, w = 500 mm = 500/1000 = 0.5 m
Let water depth in channel, d = 0.056/0.5 = 0.11 m
Hydraulic radius R = area/wetted perimeter
∴ R = area/(2d + w) = 0.056/(0.5 + 2 × 0.11)
∴ R = 0.056/(0.722) = 0.077 m
For smooth cement K = 60 (Table 11.4)
From equation 11.8:
$1.5 = 60 \times 0.077^{0.66} \times S^{0.5}$
∴ $1.5 = 11.098 \times S^{0.5}$
∴ $S = (1.5/11.098)^2 = 0.018$ m/m = **16 mm/m length**

SAQ 11.12 Flow = 75 m^3/h = 0.021 m^3/s
From Fig 4.3 N = 6; $v_1 = v_2 = 0.021/(0.5 \times 1.0) = 0.042$ m/s
Loss over baffles = $h \approx (1/2g) \times \{N v_1^2 + (N-1) v_2^2\}$
$= 0.05 \times (6 \times 0.042^2 + 5 \times 0.042^2) = 0.05 \times 0.019$
= 0.001 = **1 mm**
Loss over weirs (Fig 11.6) $Q = K (L - 0.1 \times N \times h) \times h^{1.5}$
L = 0.5 m Q = 0.021 $m^3.s^{-1}$ K = 1.838
$0.021 = 1.838 \times 0.5 \times h^{1.5}$

$H = (0.021/1.838/0.5)^{0.67} = 0.08\,m$
$= $ **80 mm loss over outlet weir**

With inlet weir set at 1 m the maximum depth of water in the contact tank is 1 m.

$\therefore$ cross sectional area of flow $= 0.5\,m \times 1.0\,m = 0.5\,m^2$
$(R = 0.5/(2 \times 1 + 0.5) = 0.2\,m$
$K = 60$
$0.021 = 60 \times 0.2^{0.66} \times S^{0.5}$
$0.021 = 20.74 \times S^{0.5}$
$S = 1.0 \times 10^{-6}\,m$
Channel length $= 21\,m$
Head loss $h = 21 \times 1.0 \times 10^{-6} = 2.1 \times 10^{-5}\,m$
$= \mathbf{2.1 \times 10^{-2}\,mm}$

Difference in height between weir levels = total headloss = 80.02 mm, so maximum level of the outlet weir is $1 - 0.08 = 0.92\,m$ above floor level

SAQ 11.13 $\dfrac{H}{l} = 5\dfrac{\mu}{\rho}\dfrac{v_a}{g}\dfrac{(1-\varepsilon)^2}{\varepsilon^3}\left(\dfrac{6}{d}\right)^2$

$l = 1000\,mm = 1000/1000 = 1\,m$
$\rho = 1000\,kg/m^3$
$d = 1.2\,mm = 0.0012\,m$
$v_a = 7\,m/h = 7/3600 = 0.0019\,m/s$
$\mu = 0.001\,kg/m/s$
$\varepsilon = 0.45$
$H = 1 \times 5 \times (0.001/1000) \times (0.0019/9.81) \times (0.55^2/0.45^3) \times (6/0.0012)^2$
$H = 1 \times 5 \times 1 \times 10^{-6} \times 0.00019 \times 3.32 \times 2.5 \times 10^7$
$H = 0.08\,m = 80\,mm$

SAQ 11.14 $NPSH = P_0 + h_0 - h_f - v^2/2g - P_v$
$P_0 = 10.2\,m$ (atmospheric pressure)
$H_0 = -1.5\,m$
$H_f = 0.3\,m$
flow $= 100\,m^3/h = 100/3600 = 0.027\,m^3/s$
pipe diameter $= 200/1000 = 0.2\,m$
area $= \pi d^2/4 = 3.14 \times 0.2^2 \times 0.25 = 0.031\,m^2$
$v = 0.027/0.031 = 0.86\,m/s$
$P_v = 4.73\,m$ (Table 11.8)
$NPSH = 10.2 - 1.2 - 0.3 - 0.86^2/(2 \times 9.81) - 4.73 = 3.9\,m$

SAQ 11.15 $E = (Q.\rho.g.h)/1000.\eta$ kW
$Q = 0.56\,m^3/s$
$\rho = 1000\,kg/m^3$
$g = 9.81\,m/s^2$
$H = 5\,m$
$\eta = 75\% = 75/100 = 0.75$
$P = 0.56 \times 1000 \times 9.81 \times 5/(1000 \times 0.75) = 36.6\,kW$
The nearest standard sized motor would be 40 kW

SAQ 11.16 Static head = 18 m
Take a range of flows from 0–100 m^3/h and calculate head loss in pipework and fittings using the method in SAQ 4.1
Calculate the headloss in the filters. Filter area $= 5.3\,m^2$ so approach velocity (m/h) = flow in m^3/h divided by 2 (half flow to each filter) divided by $5.3\,m^2$.

Calculate headloss in anthracite using Equation 11.32. Calculate headloss in sand using Equation 11.32. Total filter headloss is sum of these two:

$$1.02^{(15-10)} \times (Q/2/2.6) \times (0.3 \times 0.01 + 0.3 \times 0.06)$$
$$= Q \times 0.004\,\text{m}$$

Calculated headlosses:

Flow	Static	friction	filter	Total
0	18	0	0.0	18.0
20	18	0.2	0.1	18.3
40	18	0.8	0.2	19.0
60	18	1.7	0.3	20.0
80	18	3.1	0.4	21.5
100	18	4.8	0.4	23.2

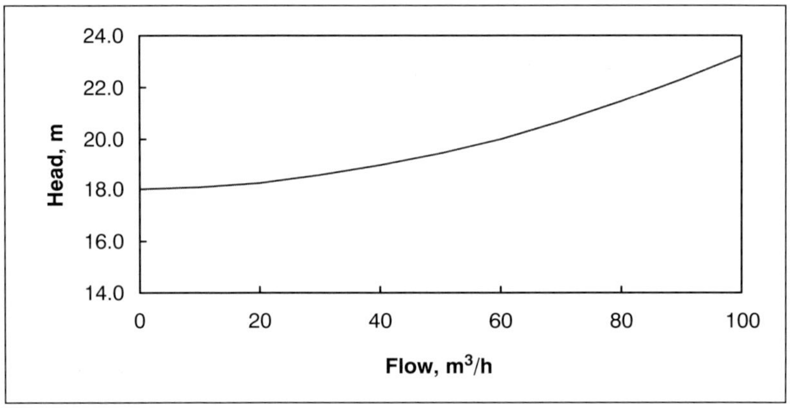

Solutions to Self Assessment Questions

Unit 12 – Particle Settlement

SAQ 12.1 Ex 1 $\mathbf{Re} = \dfrac{\rho d v_t}{\mu}$

$\rho = 999.25\,kg.m^{-3}$ (Table 12.1)
$\mu = 0.656 \times 10^{-3}\,kg.m^{-1}.s^{-1}$ (Table 12.1)
$d = 0.4\,mm = 0.4 \times 10^{-3}\,m$
$v_t = 2\,m/h = 2/3600 = 0.56 \times 10^{-3}\,m.s^{-1}$
$Re = 999.25 \times 0.4 \times 10^{-3} \times 0.56 \times 10^{-3}/0.656 \times 10^{-3}$
$= 0.34$

SAQ 12.2 $v_t = g\dfrac{(\rho_s - \rho)d^2}{18\mu}$

$d = 450\,\mu m = 0.45 \times 10^{-3}\,m$; $g = 9.81\,m.s^{-2}$;
$\rho_s = 1005\,kg.m^{-3}$
$\mu = 10^{-3}\,kg.m^{-1}.s^{-1}$; $\rho = 1000\,kg.m^{-3}$
$v_t = 9.81 \times (1005 - 1000) \times (0.45 \times 10^{-3})^2/18/10^{-3}$
$= 0.55 \times 10^{-3}\,m/s = 2\,m/h$

SAQ 12.3 $v_t = 0.55 \times 10^{-3}\,m/s = 2\,m/h$
So, settling time in 3 m tank = 3/2 = **1.5 h**

SAQ 12.4 $v_t = g\dfrac{(\rho_s - \rho)d^2}{18\mu}$ for laminar flow (eq. 12.14)

$v_t = \left(\dfrac{3.33g(\rho_s - \rho)d}{\rho}\right)^{1/2}$ for turbulent flow (eq 12.16)

So, ratio given by:
$1/(18 \times 3.33^{0.5}) \times (\rho^{0.5}/\mu) \times (g\Delta\rho d^2/(g^{0.5}\Delta\rho^{0.5}d^{0.5}))$
At 20°C density is ~1000 and viscosity ~0.001 (Table 5.1, E5) in SI unit; $g = 9.81\,m.s^{-2}$
So, ratio $= 0.0304 \times (1000^{0.5}/0.001) \times (9.81^{0.5}\Delta\rho^{0.5}d^{1.5})$
$= \mathbf{3010\,\Delta\rho^{0.5}d^{1.5}}$

SAQ 12.5 $v_t = g\dfrac{(\rho_s - \rho)d^2}{18\mu}$

$d = 50\,\mu m = 5.0 \times 10^{-5}\,m$; $g = 9.81\,m.s^{-2}$; $\rho_s = 2500\,kg.m^{-3}$
$\mu = 10^{-3}\,kg.m^{-1}.s^{-1}$; $\rho = 1000\,kg.m^{-3}$
At 80% porosity, ave density $= 0.8 \times 1000 + 0.2 \times 2500$
$= 1300$
So, $v_t = (9.81 \times (1300 - 1000) \times (5.0 \times 10^{-5})^2)/(18 \times 1.00 \times 10^{-3})$
$= 4.1 \times 10^{-4}\,m/s$
So, settling time in 3 m tank $= 3/4.1 \times 10^{-4} = 7340$ s, ~**2 h**